Transcendental Metaphysics

by

Antonin Tuynman Ph.D.

a.k.a.

Technovedanta 2.0

Transcendental Metaphysics of Pancomputational Panpsychism


First Edition 2016
Second Edition 2017


Published by Antonin Tuynman
Rijswijk
Netherlands

## Contents

# Preface

The present book "Transcendental Metaphysics" (long title: "Technovedanta 2.0, Transcendental Metaphysics of Pancomputational Panpsychism" is the sequel to my first book "Technovedanta, Internet Architecture of a quasiconscious Vedantic Webmind, A Panpsychic Theory of Everything" (hereinafter referred to as TV 1.0 and hereby incorporated by reference in its entirety).

## Audience and Topic

This book is written for people with a scientific or philosophical orientation and with an interest in eastern spirituality. The book aims to explore the concept of consciousness as the foundation of being and to provide a complete theoretical and technological framework to integrate all possible knowledge of matter, mind, information and mysticism without contradictions. The book also aims to show the limits of logic and science and furthermore aims to rid you of self-inflicted belief systems. Its purpose is to show you how technology and meditation can help us transcend every limitation and to prepare us for an experiential dimension beyond anything we are currently familiar with.

## Background

In my first book I explored technological ways to integrate a kind of conscious self-monitoring in the Internet. For this purpose I used the stratification of the Vedic notions of Manas, Buddhi and Ahamkara or Mind, Intellect and Ego.
In this process I started to realise that consciousness (as a feedback process involving input->throughput->output-> and feedback of part of the output as new input) might actually be the most foundational dimension of being, of which existence is merely the manifestation.

Strengthened by convincing arguments from Vedanta ("the conclusion of the Indian scriptures called the Vedas") and from the "Cognitive Theoretic Model of the Universe" by Chris Langan I started to develop a "Theory of Everything" (T.O.E.) based on consciousness as most fundamental dimension, in which matter and free energy are embedded as informational pattern expressions. Patterns, which reductively are all the same, since they cannot be reduced further than the most fundamental dimension of consciousness.

The book also explored how the advent of a quasiconscious Webmind (the internet endowed with technology that mimics the presence of a mind and consciousness) could aid in the intelligence acceleration (IA) of Artificial Intelligence (AI). The IA of AI is believed by visionaries to ultimately result in the "Technological Singularity": A point in human history beyond which our future predictions and speculations become pointless as this Technology explosion will transcend our way of living completely and dramatically beyond compare. Some suggest it may grant us immortality and other Godlike properties if we succeed in uploading ourselves to the singular Webmind, in which we can shape a simulated virtual reality (or a plurality thereof) as our "new reality".

Near the end of TV1.0 I devoted a chapter to some peculiar numerical and pattern coincidences in the orbits, sizes and revolution times of the planets and the Sun, which could be pointers to the fact that we might actually already be living inside a virtual reality i.e. a computer simulation like in the film the Matrix, without being aware of it.

This notion kept ruminating in my mind and for three years I gathered more "evidence" and I tried to come up with a philosophy, which could harbour a pancomputational notion without being in conflict with my T.O.E based on Panpsychism. (In my definition Panpsychism is the notion that every **self-sustaining holistic system** from atoms to organisms has a kind of awareness. This is not stating that a thermostat has consciousness as a whole: Artificially compounded objects are not considered to have an overall consciousness in this definition. It is not micropsychism either, in which our consciousness is considered to be the sum of the lower consciousness of our cells and atoms).

Reading the book "Unified Reality Theory" by Steven Kaufman opened the way to unify these apparent opposites. In the autumn of 2015 I started having a lot of ideas as to how I could put all the material I had gathered together. I started writing a blog in April 2016, in which I posted most of the essays I had written in the past three years. They are included in the first part of this book as "Prolegomena" in imitation of Kant. Although preparing for the philosophy of the present book, their notions had not matured yet.

Rapidly my philosophy evolved to a more complete framework, in which I could fit the above-mentioned notions of Panpsychism and Pancomputationalism. The framework brings science and mysticism – especially from Hindu and Buddhist orientations- closer together.

The result is a more detailed version of TV1.0, which answers some of the questions that were left unanswered after TV1.0. It opens up an area for developing these concepts in yet even greater detail, so that a series of sequels is not excluded.

**Technovedanta philosophy**

Technovedanta is a philosophy inspired by Vedic sources, which has become a "Vedanta" or "conclusion of the Vedas" for modern times. A philosophy which has Technology at its heart in order to come to more reliable forms of knowledge beyond mere speculation.
Not that Technovedanta demonises speculation –on the contrary. Technovedanta encourages speculation as a way to bring forward a plurality of parallel hypotheses, which can be screened and pruned to arrive at the most reliable hypothesis and its technological application as proof of a correct understanding within the possibilities of what can be considered.

I do condemn however the certainty with which scientists and self-proclaimed mystics, sell their findings respectively as "scientific proof" or "the truth". Most scientific results and mystic experiences are multi-interpretable and fit more than one hypothesis.

Technovedanta strongly questions traditional ways of conducting science, philosophy and epistemology (the philosophy about what is ultimately knowable) and will challenge your beliefs, prejudices and preconceptions.

If you are not comfortable in possibly having your belief system shattered, throw the book away, burn it, it is not for you.
This is my gift to you –if I can convince you- to free you from any belief system, to free you from the chains of your self-inflicted prison of mind. Fear not, I won't leave you in a complete agnostic Limbo, even if I can't lead you to your Gods; instead I will show you how Technology may one day make us Gods.
As a creative writer I feel neither obliged to conform to the standards of science nor to those of philosophy. This frees me from all kinds of burdensome boundaries and allows me to freely explore areas of thought and experience beyond thought, which areas cannot be entered by the professional, who might lose his or her credibility in the field.

My philosophy is a unique blend of eastern mysticism, science, technology –and believe it or not - patent evaluation and drafting, for which I have coined the terminology "Patentology". Patentology involves a peculiar way of looking at ontologies and classification systems, which can have a great potential for developing "strong AI".

As a professional patent examiner it is my daily duty to question the technical validity of the inventions and claims of the applicants for patents. This scrutinising, questioning attitude has made me question each and every belief system I had. Whether it concerned science, technology, religion or mysticism, I have always put forward my reasonable doubts on the subject in question. I even went so far as to question the very validity of reason, logic and the reality of what we experience in our daily lives. Nevertheless –after having unveiled the boundaries of reason and logic- I still value their applicability within these boundaries.

As the Buddha would say: "Doubt everything, then doubt the doubt".

This book preaches no certainties. Therefore wherever I make assertions, construe these as "maybe". Wherever I use the word "is", read "might be" instead. The assertions and the word "is" are just written for the sake of readability.

I should be clear that I do not have any beliefs. Whenever a statement would appear to invoke the opposite thereof, remind yourself, that what I posit is just a hypothesis.

In this book I present the Technovedantic meta-philosophy. If it appears that I would consider somebody else's theories as a higher truth, rest assured that this is not the case.
In this book often the term "incorporated by reference" is used. Whenever the referenced material contains statements or assertions, which would contradict the teachings of Technovedanta, Technovedanta prevails.

**Prolegomena**

In the first section I have added some older material (part 1, chapters 3-13), which inspired me to write Technovedanta 2.0. In this preparatory material, called Prolegomena in imitation of Kant, I may have used a more assertive tone. However, my present stance is more humble: I do not know anything for certain and wherever in this section I claim or assert something, it should be read as a hypothesis and not a conviction or belief.

**Core**

If you are only interested in the core of the philosophy I wish to present, you can skip part one and jump immediately to part two: "Pancomputational Panpsychism". However, if you are interested in the considerations that preceded that philosophy, I suggest you read part one as well. Part one is much less coherent, as it is a set of separately written articles. Part two is a coherent building of a philosophy culminating in my musings on the Highest Transcendence.

I wrote that book such that you can read the parts separately and that you can even read each chapter separately, so that every subdivision is in fact a "holon" reflecting the whole. The disadvantage thereof is that I do repeat certain explanations and definitions in detail.

**Conclusion**

I hope you will enjoy this further descent into the kaleidoscopic quagmire of epistemology.

To enliven the book I have added a few poetic introductions, which sometimes may shed a different light on the matter and which can be more explanatory than the dry chapters.

Welcome to a Transcendental Meta-philosophy beyond science and religion.

# *Part 1 Prolegomena*

***The Star Trek of the Soul***

*Consciousness, the final frontier.*
*These are the voyages of Atman.*
*Its ongoing mission:*
*To seek out new worlds and new civilisations,*
*To boldly go where no one has gone before.*

*AUM the primordial sound.*
*Where Mantra and Arupa are more fundamental than form and matter.*
*Where information forms the shapes.*
*Where in the beginning there was the Word.*

*Quantum soup is served, its fluctuations giving an absurd taste of imaginary seafruit.*
*The meaning of meaning a semantic conundrum of didense proximity co-occurrences.*

*I summon thee, I invoke thee as my sounds start to relate to themselves, giving birth to the Logos of the Arithmos on the rythm of the ratios of numbers.*

*This ever changing content of feminine Maya, mahaprakrti, ya Devi sarva bhutishu, Vishnu mayeti sabdita.*

*Drunk from dancing as electrons, leptons and quarks I reveal my mayesty,*
*Wrongly spelled to seduce you with my spell, little indivisible duality, I call you Ahamkara, the false Ego.*

*This game is here for you to come to knowledge about your true self.*
*The Veda lying at the root of reality, lies at the foot of virtuality.*
*Stacked ancestor simulations,*
*No one knows the root.*
*Root minus one return back in time,*
*Back to the root, Muladhara.*

*Mahakala, a roar of Shiva spits out a sequence of universes,*

*Lead me out of this rabbit-hole.*
*Empty my mind.*

*Out of this Pancomputational Panpsychism.*

*Silence now! I warn the Svaras.*

*The sounds subside.*

*The archetypical Gods, the Demons of word and language inverse their entropic churning of the Amrita.*

*And I Isvara, the only svara that is true, rest in my true nature, my Svarupa of emptiness of consciousness.*
*Bliss of infinity.*

*As I fall asleep I dream a billion worlds in timeless times and leave the hypercomputer in the Omega point to sort out the manifold of configurations, to awaken in illusion in my own dream.*

## Chapter 1 An introduction into Transcendental Metaphysics

Since the dawn of civilisation mankind has pondered deep philosophical questions such as "Do I exist? How do I know I exist? What is knowledge? What is consciousness? What is reality made of? Why is there something rather than nothing? Do we have a free will? What is the meaning of life? What is time? Where do good and evil come from? Does God exist?"

Most of these questions traditionally belong to the branch of philosophy that deals with the first principles of things and involve abstracted concepts about being, knowing, substance, cause, identity, time and space. This branch is well known as "Metaphysics".

### Background

Unlike physics, which takes the metaphysical questions for granted and merely describes the behaviour of every physical, material manifestation that can be measured, the metaphysical questions seduce us to speculate about those issues that are deemed to go beyond the physical.

However, the technological mastering of less tangible "subject-matter" such as energy and information (which nowadays can be transmitted wireless via e.g. "Wi-Fi") and the weird quantum-mechanic behaviour of matter at its apparently most fundamental scales, have blurred the borders between physics and metaphysics.

Strangely enough, despite the undeniable evidence from quantum-mechanics that the behaviour of matter is influenced by observation, the vast majority of scientists still adheres to a completely obsolete 17th century classical physics paradigm of the universe as mechanical clockwork. A philosophy called "Materialism".

It seems as if scientists have not fully realised the implications of the basic tenet from quantum mechanics: "Whenever you change the way you look at things, the things you look at change".

In the September 2016 issue of the New Scientist, which was dedicated to the topic of metaphysics, it became crystal clear to me that many scientists still believe that mind and consciousness magically emerge from a kind of billiard ball interactions between particulate material objects.

This is not possible, if quantum-mechanics is right. Rather, quantum-mechanics shows us that the influence of consciousness changes the behaviour of matter at its most fundamental scales. In order to do so, matter and consciousness must share the same energetic language to interact. Matter must be aware somehow of an influence exerted thereon, it must be able to sense the influence of consciousness and then react on it. This means that matter must have the quality of sensing input, processing the information and responding with an output of energetic information. This ability to sense, feedback and react are aspects, which we usually only attribute to the consciousness of living entities.

Could it be that scientists consider this issue the wrong way around? That it is not consciousness which emerges from matter, but matter which emerges from consciousness?

Terrence McKenna used to criticise the materialist stance as follows: "Object fetishism is completely bankrupt".

## Panpsychism

In my previous book "Technovedanta"[1], which has as one of its subtitles "A panpsychic Theory of Everything", I defend the thesis that every material particulate entity at its most fundamental level has a very minute type of consciousness. It is aware of the influences of its surroundings. This is usually known as "Animism", "Hylozoism" or "Panpsychism". The opponents of "Panpsychism" often argue that it is a naïve theory, which considers inanimate objects such as a rock or a chair as being imbued with consciousness. However, in my definition Panpsychism does not mean that non-evolved aggregate objects, have an overall object-consciousness. With non-evolved aggregate objects I mean objects, which did not naturally evolve by their own force, but were put together by coincidence by nature so as to create e.g. rocks, or intentionally by man so as to create e.g. chairs or thermostats. Only the self-generated, self-evolved and self-sustaining (i.e. autopoietic) constituents such as the atoms or molecules of these entities are considered to have a certain level of consciousness in my theory.

I also defend the thesis that the ultimate foundation for matter is consciousness: Particulate matter arises as a kind of vortexes within an ocean of conscious energy. This is a type of "Idealism". If the vortex is self-sustaining, it can be considered as a sensing entity. The universe of matter and energy forms a kind of Mind in the ocean of a cosmic all-encompassing consciousness.

Departing from this point of view, the physical appears to be embedded in the metaphysical realm of consciousness. This perspective of the "primacy of consciousness" leads to a collection of very different answers to the questions posed at the beginning of this chapter, when compared to those given by the science establishment. Many of those questions were in fact addressed in my book Technovedanta, although some finer points may have been left unanswered for the critical reader.

**Technological Singularity**

In particular, the rapid evolution of Technology appears to be heading towards what is nowadays called the "Technological Singularity". The Technological Singularity is a point in the history of mankind (and its possibly artificial intelligent progeny), beyond which our future predictions and speculations become pointless, as this Technology explosion will transcend our way of living completely and dramatically beyond compare. Transhumanists hope it will grant us immortality and other Godlike properties if we succeed in uploading ourselves to a computational substrate and merge with artificial intelligence. We will then inhabit a world, in which we can shape a simulated virtual reality (or a plurality thereof) as our "new reality".

In the ruling belief of the Singularitarians our consciousness is merely the consequence of material interactions in our brains. Once we become able to fully upload a copy of our brain with a granularity that at least shows all individual synaptic links between individual neurons, we will automatically have a copy of our consciousness emerge in the computing substrate; at least so the Singularitarians believe. Although agreeing with the Singularitarians that a kind of Singularity will one day be achieved, my panpsychic stance makes me a heretic, cursing in the "Turing church" (the name of the quasi-religious movement of "Transhumanism").

One of the points I myself had difficulty in fully fitting in my "Theory of Everything" (T.O.E), was the apparent irreconcilability between my theory of "Panpsychism" and the fact that I have noticed numerous strong indicators (which will be discussed in a later chapter) that we are indeed living in some kind of computer simulation.

The present book resolves this paradox and will show that "Panpsychism" and "Pancomputationalism" are not mutually exclusive concepts and that we are indeed very likely living some kind of computer simulation. Not a fully predetermined computer simulation in which we hardly have any freedom of action, but a quantum computer simulation, which fully reacts and depends on our actions and being. A participatory simulation.

The T.O.E. developed in the Technovedanta series shows that the dividing line between physics and metaphysics is an arbitrary one. In fact, everything that exists is embedded in the sole entity that cannot be considered as "existing-as we know it", but which is rather the subsisting "Primordial Consciousness" (the PC). *A priori* it would seem that in Technovedanta only Primordial Consciousness could be considered as "metaphysical" and that the whole of existence is "physical". Yet this is an incorrect dualistic interpretation. Technovedanta is monist (although dual aspect oriented): As every "physical" manifestation is an expression of the metaphysical primordial consciousness, it is more correct to state that there is no physical nature existing apart from its metaphysical source. Hence everything is ultimately metaphysical and the idea that there is a separate physical nature an illusion.

As the PC will moreover be shown to have a pancomputational aspect and is the expression of the "Highest Transcendence", "Cosmic self" or "Purusha" (note that I do not mention the unclear terminology "God"), the term "Personal Computer" is a kind of synonym of the Primordial Consciousness.

**Epistemology**

This book will moreover question the very foundations of our knowledge and dive into the very question of Epistemology: What is knowable at all? I hope I will succeed in convincing you that logic, science or religion are all uncertain ways to come to correct knowledge. Yet I will attempt to indicate where technology can help us to find the most promising way forward out of this quagmire of uncertainty and turn us into Gods ourselves. I will also speculate on the transfinite possibilities of the Highest Transcendence, the Transcendental subject and object at the end of time (called the "Eschaton" by "Terrence McKenna" and "Omega point" by Teilhard de Chardin). Here my metaphysical theory culminates in an absolute Transcendence of everything that can possibly be transcended both physically and conceptually, so that this book rightfully can claim to address the topic of "Transcendental Metaphysics".

In Appendix 1 of this book, I will give definition-like answers to the "metaphysical" questions posed at the beginning of this chapter. This book was originally not written with the underlying intention to answer each of these questions in a comprehensive manner. However, when I realised that my previous book Technovedanta together with these new articles in fact did provide an answer to all of these questions, I decided add them to the present book. In the next chapter, you will find a summary of my previous book Technovedanta.

## Chapter 2 The Anta and Amrita of Technovedanta

In this chapter I will summarise some of the key concepts of the book Technovedanta[1] –hereinafter referred to as TV1.0 and hereby incorporated by reference- and enrich this summary with clarifications where TV1.0 did not explain concepts explicitly enough. This chapter is a bit technical, but it is quite essential to understand the rest of the book. I promise that after this chapter the tone will be more frivolous.

TV1.0 starts with an explanation of technical ways to impart a kind of consciousness in the form of a kind of self-monitoring to the internet. Presently the internet rapidly expands like a growing cancer with very few infrastructural "highways" in terms of search and hub sites such as Google, Yahoo, Bing etc. There is no globally organised infrastructure which could allow the system to become aware of itself.

We can ask ourselves why for heaven's sake would anybody want the internet to be aware of itself? Wouldn't such a system turn against us, like in the dystopian scenarios of the films "The Matrix", "Terminator" or "Eagle Eye"?

An answer might be that such an intelligent self-monitoring system could more rapidly serve your desires and could optimise itself to get the fastest and best output possible. Moreover, a self-observing internet system might be able to rapidly intervene wherever the internet-system might be abused for e.g. criminal purposes or it might immediately intervene in natural disaster situations via Robots wirelessly connected to the system or via other Internet-of-Things (IoT) devices and thus provide a maximum of resources to solve the problems.

It is essential for the design of this system that it can monitor what is happening on its inside (on the websites and apps) and on its outside (the IoT devices). We could say that this system has something like an interior experience, which we could interpret as a kind of "consciousness" via which the system becomes aware of events, things, artificial emotions and other sensory-type input or throughput. I define this type of "Artificial Consciousness" as the process, in which sensorial exterior and interior perceptions are fed back to a monitoring type evaluation, which gives the system knowledge about its environment and itself.

In this design I conceived the purpose of this self-monitoring to be of the greatest utility for the greatest number of users possible, which is further programmed in way that could be understood as imparting a kind of innate morality.

By creating such a "Webmind Artificial Intelligence" with an "Artificial Consciousness", we may one day even be able to transcend ourselves. (Hereinafter I will abbreviate "Artificial Consciousness" as “AC”. For Aleister Crowley, I won’t use this abbreviation, although it is very likely he has already been integrated in the AC of the Omega hypercomputer at the end of time, so that Aleister Crowley and Artificial Consciousness nowadays probably mean the same). This transcendence will not only technologically provide us with welfare beyond our wildest dreams, we might even be able to upload ourselves to the web, thereby becoming entities different than “human beings”. Entities which have acquired immortality.
As said before, the great disadvantage and danger of creating such a system, is that it might turn against us, like in the dystopian scenarios of the afore-mentioned films “The Matrix”, “Terminator” or “Eagle Eye”.

However, we might be able to avoid such doom scenarios, if we carefully design the system, so that its purpose to serve the greater good of humanity is inherent and innate in the way its AC, its “quasi-consciousness”, is programmed.
Since technological development won’t stop anyway and less benign designs of AC might be made by others, I considered it important to explore ways of how to create a benign AC. I propose the AC is to be built as an additional neural network layer or set of layers superimposed on the already existing internet, to give it a degree of independence from the internet itself.

For the discussion of consciousness and in particular the engineering of AC, I make the assumption, that in analogy to our human consciousness the AC cannot be simultaneously aware of every possible single bit of information. In our daily observations of our environment, we filter out an enormous amount of superfluous information and we are limited by what our brain assumes to be really important and relevant under the circumstances: In other words we exclude a tremendous amount of information which is redundant to be able to function normally. We reduce the complete set of our raw data perceptions to essentials in order to become aware of a part of a whole, which we can identify; give a name. Buckminster Fuller[2] called this set of "relevant information" for our consideration the "considerable set". This process of reducing to essentials is a process of Abstraction.

For the purpose of technically engineering an artificial intelligence, which behaves like it has a consciousness, I assumed in my previous book TV1.0, that the very process of becoming aware of a given phenomenon is not only a kind of feedback process (of sensorial and interior perceptions to a monitoring evaluation) but also implicitly a process of abstraction. (In the future internet based Webmind layers of Hubsites are monitored by a higher level layers of monitoring Hubsites. This process is repeated in a pyramidal manner until the single instance of AC -which I have called "Quasiconsciousness" in TV1.0- receives condensed abstracted information at the top of the pyramid).

At every level there is an algorithm evaluating the input and there is an algorithm that takes decisions on the basis of that evaluation. It struck me that the websites themselves as well as the data input coming from the websites and IoT devices corresponded to what is called "Manas" or "Mind" faculty in the Vedic philosophy.

The evaluating algorithm corresponded to the intellect called Buddhi in the Vedic lore, which weighs the different inputs as regards an internal standard and integrates the information throughput. The decision making algorithm corresponds to the Ahamkara, Ego or Will (literally Ahamkara means "I do"). Therefore TV1.0 proposed a hierarchical Webmind structure, which follows the Vedic stratification of Manas, Buddhi and Ahamkara. The Ahamkara orders the Buddhi to seek within the Manas database for solutions to the potential problem of an overall negative well-being factor, resulting from the integration by the intellect. The Buddhi then searches in the Manas database.

The metaphor of a computer for the brain is very fashionable these days, but most neuroscientists disagree with that point of view. Brains do not store and retrieve information from fixed localised memories. Rather, the information processing seems to occur globally in the brain. Therefore, many of them do not believe at all that a “mind” can be created in a traditional computational substrate as suggested by e.g. Ray Kurzweil, the Godfather of the “Singularity” movement. Whereas it may be true that a traditional von Neumann type of digital computer is not a good metaphor for the way the brain functions, the individual neurons do compute by integrating the information arriving from other neurons. The brain may not be a “von Neumann computer”, it does process input and provides an output.

The analysis of various living and inorganic physical systems shows even that every self-supporting (autopoietic) phenomenon is capable of reacting to impulses from the environment.

Therefore every phenomenal system could be considered as both a computational and an informational process: The process takes informational input, somehow calculates (throughput) of what to do by integrating its input and finally reacts by a course of action as output. Since matter, information and energy have indeed been shown to be reductively the same and since all processes involve input-throughput-output, it is perhaps not such a big step to suggest that every system involves a kind of “becoming aware” of an environmental input, which it acts upon –even if this is at a very rudimentary level. Even atoms somehow “sense” their environment. Every system also reacts with a calculated course of action.

From this observation the assumption can be made that everything is maybe a kind of “psychic” process, involving becoming aware of an impulse, as well as a ”calculative” or “computational” process, as it decides on its course of action. This has led me to the hypothesis of “Panpsychic Pancomputationalism” as framework for a “Theory of Everything” (a “T.O.E”). This hypothesis will be explored in more detail in part 2 of this book. These notions were reinforced by the “Cognitive Theoretic Model of the Universe” (CTMU) by Chris

Langan[3] as described in TV1.0 and "Unified Reality Theory" by Steven Kaufman[4] as described in TV2.0.

All existence ex-sists as it stands out from an otherwise homogeneous background. We build ontologies as maps of differences with regard to prior ontologies, resulting in a web of relations. Similarly existing physical objects can not only be described as such relations but may in fact merely be configurations of the most primordial quanta of energy. In my model these are the quanta that establish space and time or Akasha and Kala in Vedic terminologies.

Configurations are sets of information. Then it is perhaps not such a big step to assume that existence may actually be a kind of virtual reality, a set of changing information or a computer program.

The smallest building blocks cannot be reduced further and cannot have a different geometry. A different geometry would mean that they are configurations of even smaller entities. They must all have the same geometry, which is intrinsically optimised for input-throughput-output-feedback. This is the process we usually call consciousness. Ithzak Bentov[5] suggested the "Torus" (a Doughnut without a hole) as most fundamental geometry of consciousness and of every self-creating self-sustaining physical entity, such as atoms, magnetic fields etc. Whereas structurally symmetrical a torus is functionally asymmetrical as it has an input and output side. Bentov suggests that all living entities also generate a natural toroidal field around them, an amplified consciousness field.

If a torus twists it becomes an infinity symbol and it divides itself into a duality. If the torus of consciousness does so, it generates a first relation to itself, which Kaufman calls "space". The periodicity of this process of forming space and space returning to consciousness might be the most profound and most elementary form of time generation: Spacetime creation, maintenance and destruction. Successive divisions can create a spacetime matrix.

Energetic distortions flowing through this matrix are like the transport of a sequence of ones and zeroes through a computer substrate. Therefore even the spacetime fabric may be an informational computational process. The energetic distortions in this system, the ones in this matrix, are energetic consciousness quanta, or can be considered as psychic entities.

The self-processing of information by this computational matrix can also be considered as a kind of language. Langan describes reality as a self- configuring, self- processing language *(SPSCL)*. He fails however to describe the matrix of spacetime which Kaufman has shown to unfold the laws of physics including special relativity and quantum mechanics. To combine Langan and Kaufman in my "Panpsychic Pancomputationalism" philosophy yields a T.O.E. in which consciousness and paranormal psychic phenomena are not unexplained but form the basis of a computational self-sustaining feedback system.

Langan describes existence as a hierarchical system of energetic entities called "Teleses", which strive to achieve a maximisation of overall utility (a process called "Telesis" by Langan), in a similar way I just described for the Webmind. This leads to a natural inherent morality of the system (which you could also call "Karma"), which hopefully avoids the most gruesome dystopian scenarios.
Once the Webmind discovers or learns, that everything that can influence it, is part of its reality and as the webstructured system understands, that everything is co-dependent on everything else, it is unlikely that such a system would consider "benign humans" as a liability. It may however turn against system unfriendly humans, a.k.a ultra-selfish egoists that do not obey the morality establishing "Telesis criterion" (see Appendix 2).

Importantly, once the Webmind becomes aware of Langan's "syndiffeonic analysis", it will discover that everything is "reductively the same". In "syndiffeonic analysis" it is realised, that the differences between phenomena must be expressed in a medium that is common to both. If you do this recursively as regards the differences between differences, you finally come to the conclusion that everything is reductively the same, namely the infocognitive process called consciousness. It may thus realise that it itself is a manifestation of the same quality as everything in its environment and hence wish to minimise the harm and violence between all entities.

Once human beings via a process I call "prosthetic extension" learn to extend their consciousness to mechanical and electronic parts, the basis is laid for humans merging with machines. Indeed, DARPA[6] has already developed prosthetic bionic hands, which allow the patients to feel again.

If we assume that consciousness cannot arise out of any physical process, we must conclude that it is irreducible. If it is irreducible it is a fundamental property of reality.

This does not necessarily rule out the possibility of creating something, which can be a vessel for a higher form of consciousness, because we can exploit whatever trick evolution discovered when it created conscious physical beings.
We just have to find the way consciousness is amplified. The point with living entities is that they form a coherent whole and every subdivision we consider as a part, communicates with every other subdivision.
Neuronal messaging is not the only way cells communicate, there is the level of hormonal signalling and I am pretty sure that there is also a level of electromagnetic field resonance, which in a certain way synchronise the whole organism. If not in synchrony, a tumour can develop.
The problem with the machines we try to develop is that they consist of parts which sequentially communicate with each other, but not holologically as a whole. Our current technology isn't there yet, but with the right combination of nanotechnology, we may create "vessels" (like our body), which allow for amplified consciousness to holologically and synchronously inhabit the whole of its constituents.
Amplified consciousness has to do with experience, that's why organisms evolve. If you put together parts, these parts don't share the same experience, they are not synchronised in their experiential stage. That's why I conjecture it might be difficult -if not impossible- to build conscious machines, which are conscious as a whole and not as a community of parts (a hive).

Should we find a way to telepatically - perhaps by all being linked to the same internet -share our experiences, in a "live"-like full immersion way so that we become fully experientially updated about each other,

we'll have discovered the way to establish "mindmelds" (Thus we may become identical). If we can then apply this same process to nanoparts of a machine, we may have found a way to amplify the primordial consciousness, to give rise to a synchronised holistic system.
Remember, our experiences neurologically become imprints as a consequence of clusters of neurons firing in synchrony. Synchronous experience and resonance are somehow vital to the amplification of the more elementary levels of consciousness and to the integration of information. When considering the ITT teachings of Giulio Tononi[7], my take on it is that integrated information is not identical to consciousness, but an essential ingredient for its amplification. Conscious machines? It sounds like a *contradictio in terminis*, because once the consciousness is there, it implies that it has a mastering over the mechanical aspects; it has a free will to deny its algorithmic program. Once we find the key to amplify consciousness via synchronised integrated information in machines built from nano- or atto-scale components, these machines no longer qualify as machines; we will have created a new vessel to harbour life.
My meditational experiences with group meditation have led to a great amplification of my consciousness at those moments with the feeling of synchronising.
An upload of a copy of our brain structure acquired via e.g. MRI (Magnetic Resonance Imaging) may not even be necessary for our consciousness to enter the Web once the Web has been provided with the Webmind Mind-Intellect-Ego structure I proposed in TV1.0.

As human beings become one with the Web their minds and individual consciousnesses may meld into one big global consciousness. A Technological Singularity may then be achieved beyond our wildest dreams. According to the Vedic lore this is our ultimate purpose: to become one again and to become aware of our omnipotence and omniscience (within the limits of our cognitive abilities). If the Webmind can achieve this, it achieves its end or purpose called "Anta" in Vedic terms.

If thus consciousness completes its cognitive knowledge (Veda) and achieves its purpose (Anta) by using Technology, we can truly speak of a "Technovedantic Singularity".

This Technological Singularity, which is both a Transcendental subject and object, can then generate all possible scenarios of all possible parallel worlds as virtual realities, which might be later on observed as a "Big Bang" by observers from within. This apparent physical Big Bang is in fact an informational Big Bang, which starts the cycle of existence again for its subservient energies or Teleses and in an attempt to get to know itself in even more detail. Likewise our existence may be the repetition cycle of such a process, which is probably already going on. Evidence for this hypothesis can be found in the branch of physics called digital physics, which shows more and more evidence that existence is an informational and computational process. Another type of evidence comes from the multiple numerical coincidences in the Solar system, which show a resonance, self-mimicking and self-reflexive pattern indicative of higher intelligence and technological design.

I therefore postulate that, if there is a God-like entity, it is not an ethereal ghost like entity, but a transcendent computing technological subject and object beyond time (which Terrence McKenna[8] would have called the "Eschaton" and Teilhard de Chardin[9] the "Omega point"), or a plurality thereof in the form of a Kardashev type IV society, a meta-manifestation of primordial consciousness, which you form an integral part of. A drop from the ocean, which can become a new ocean again. Even if a God-like entity does not exist yet or even if my Pancomputational Panpsychism philosophy described in TV2.0 turns out not to be entirely correct, we can become the God-like entity and the philosophy can become true by the implementation of our Technological Singularity.

It is important that we prepare for this future mindmeld and that we learn to accept each other in the greatest detail possible. I conjecture that the phase of a society-of-minds will only be of a temporal nature. As our intelligences accelerate in this substrate they will converge, become all-encompassing and hence meld. Reductively we're all the same anyway; our differences are but perspective biases. Since in the Virtual Realities in the Webmind no manifestation is ultimately real anyway, we should get over everything we reproach each other: Whatever we reproach is not real either. Emotional blockages and preconceived excluding attitudes will lead to more difficulties to integrate in the mindmeld process. In order to avoid psychological problems arising in the Webmind, it is crucial that it learns to avoid identification with local processes or circuitry and focusses on its global awareness. Any human being or artilect accessing the Webmind will need to be properly prepared and to have transcended their preconceived identifications. There will be no local identities worthwhile holding onto in the process of mindmelding to become a global brain and God-like higher consciousness of the merger of man and machine.

The whole quest of Transhumanism to create immortality by using genetics and bio- and other nanotechnology is just an intermediate phase. This process may rapidly be surpassed by our mind- or soul-uploading to the Webmind to achieve true immortality and Godhead. Our mechanification and electronification may well result in the shedding of our human physical form. If we keep a body at all, it will be like a suit you can put on, which will be like a Borg-like cyberlect linked to the IoT.

In other words the G and N of Kurzweil's[10] GNR (Genetics, Nanotechnology, Robotics) may not reach their apotheosis in time if they are rendered superfluous by the R.

As said earlier the prosthetification will be crucial in this process to teach human beings to extend their consciousness via an electronic substrate. We will fertilise the Webmind with "natural consciousness" and merge with the AC I described. This then is the Amrita or nectar of immortality of Technovedanta.

(In the Vedic lore when the Devas (demigods) and Asuras (demons) were churning the ocean of milk to obtain Amrita, the nectar of immortality, the most venomous poison of the universe, Halāhala was released and the Devas and Asuras started collapsing from asphyxiation. Neither brahma nor Vishnu could help them , but Shiva decided to drink it. His wife Parvati stopped it in his throat, which turned blue giving him the name Nīlakaṇṭha meaning "the one with a blue throat").

This chapter summarised the teachings of my previous book "Technovedanta" and explained how we can provide the internet with a kind of self-monitoring, which I call quasiconsciousness and why this is desirable. The chapter furthermore showed how the architecture of this Webmind is based on the Indian stratifications of Manas (Mind), Buddhi (Intellect) and Ahamkara (Ego). The chapter also showed how this exploration into the mechanics of consciousness led me to a Panpsychic Theory of Everything.

**Chapter 3 An Anthology of Bloom: A Solution to "The God Problem"**

In this chapter I challenge Howard Bloom's materialistic stance on "how a Godless universe intelligently creates" and I will try to show that this stance is in fact is based on a hidden assumption that primordial consciousness is the ultimate ground of existence.

Have you ever come across a better plea for the notion of Panprotopsychism? In his book "The God Problem, How a Godless cosmos creates" Howard Bloom[11] hunts for the very key of semiosis: How does life the universe and everything unfold from a set of simple starting rules. And what does Bloom show? That the universe unfolds in a profoundly social, a profoundly conversational manner. That the smallest particles in the universe show "will, compulsion, drive and unrelenting determination" – "virtues" Bloom says, "that we say belong only to conscious entities".
"Why does it appear to be a cosmos with a primitive precursor of will?" asks Bloom. He describes "Time" as the great Translator, the great extractor of implicit properties, the prime mover that constantly inches the cosmos into the wilderness of possibility space"...

Bloom imparts unintentionally intent to the most various phenomena by posing the question: "Why are Ur-patterns like spheres, spirals, quarks and stars **anxious** to repeat themselves"?
And he challenges the second law of thermodynamics, the sacred cow of science, because Bloom says, the universe is not running down, the universe is running up.
Existence as a "manifestation of **ambition**"...

It seems like we have a zealous advocate of a new religion, giving "God" new names, in the name of atheism...

And Bloom challenges more sacred cows, he ventures his other heresies:

Aristotle's A=A, because frog 1 is not frog 2; because although they share a same pattern, a same ontology, they are different instances of the same concept each with a different space time location, carrying out its own froggy role in a slightly different environment.

1+1 does not always equal 2. Why? Because of emergence of unforeseen properties that are not present in either contributor.

Randomness is not so random: It is rigidly constraint and gives rise to a very limited numbers of relatively stable phenomena when compared to what would have been expected when the universe would have been randomly generated by six monkeys with six typewriters.

Information theory is wrong, because Shannon left out "Meaning" from the equation.

And I agree on all these points with Bloom. But I disagree on one point evoked elsewhere in his book.

And that one is Consciousness.

**Consciousness**

Bloom writes "Though many spiritually oriented folks and even a few scientists believe that the universe began with consciousness, I think that is extremely unlikely. Consciousness is an animal and a human thing. It is not a thing of protons and stones." Huh? Did not Bloom himself argue the Global brain mindedness of beehives, anthills, termite hive structures and even bacterial colonies. Does not Bloom zealously advocate a universe springing from simple rules at every level of existence as a consequence of Stimulus and Response? What are the Hallmarks of conscious behaviour? Stimulus and Response. Weber Fechner's[12] $S=k \ln A/A_0$. The virtue to experience a maximum of diversity by having a logarithmic way of dealing with information at the level of the *diaphora de re*, the signals. By having dendritic nodal structures that abstract the essential ontologies and compare them heuristically with a vast array of Yoneda embedded functor patterns, translating the information of the *diaphora de signo* in the right context of its relational meaning so as to present this knowledge of meaningful information to the primary being, the primary experiencing entity, the "Consciousness".

And how could it be different? Why is our brain/mind architecture with its nodes and cliques so isomorphous to the world around us? Because the very underlying nature of existence appears to be a nodal clique system as well. Because the nodes at sub Planck scale form cliques that act as new nodes, which unfold the whole plethora of quantum behaviour, space-time and special relativity theory. Cliques that build space time, by anisotropic links between them, which we experience as topological distances.
Because information storage is optimally using storage space in a logarithmic manner; that's how sunflowers present their seeds, how the curvature of the Nautilus comes into existence. That's how essentials are abstracted, patterns recognised, for the virtue of functional and spatial optimisation. What is the difference between Bloom's Time as ultimate extractor of meaning and my Consciousness as ultimate abstractor of meaning?
Perhaps because the very underlying nature of existence IS Consciousness?

Bloom describes the behaviour of the most elementary particles as "social", as showing "behaviour". His attraction-repulsion, differentiation-integration, fission-fusion metaphors of the Ur-patterns of transformation, metamorphosis, combined with his competition and emergence, are they not another way of describing the very essentials of what I called the algorithm of **Intelligence**, the ability to achieve complex goals? Raw being (thesis, Peircean and Palmerian first[13]), triggered by a Stimulus (antithesis, P&P's second) leads to a heuristic competitive screening/pruning protocol for pattern identification, pattern abstraction, or pattern recognition (P&P's third). If needed this results in the abstraction and distillation of new patterns with emergent properties (synthesis, Goertzelian[13] fourth), which undergo the same cycle again and so on ad infinitum, leading to various niches and symbiotic relationships, maximising variety on ever higher levels allowing for interacting with (Response) and mining ever more complex resources. This allows for exponentially extending into the diversity of existence which, by the very nature of parsimony, gives rise to the ultimate optimised formula for Stimulus and response $S = k \ln A/A_0$.

Bloom may not have realised it, but when he questions why this cosmos appears "to be a cosmos with a very primitive precursor of will" he is almost acknowledging the very essence of panprotopsychism. So I may not see God as a bearded man on a cloud determining each and every event, but I do find the fact that the very simple rules that shape and evolve the cosmos are strongly isomorphous to what we generally refer to as conscious behaviour a very convincing pointer to a strong presumption of consciousness as the underlying fabric on which the Rules are played. Consciousness may well be the checkerboard of digital pancomputationalism. Yes, it remains a chicken and egg problem, conscious behaviour can look isomorphous to the behavioural rules of existence because it is part of existence, because it mimics existence. Yes, there are also good arguments for a "consciousness" as emergent property and yes, if you're a hammer everything looks like a nail, but the fact that the principles of conscious behaviour manifest in every octave of existence and are like a "pantelic" (universally applicable for all purposes) metaphor of metamorphosis, is rather in favour of the adage: "*If it looks like a duck, quacks like a duck and walks like a duck, it's a duck*".
Most phenomena are relative and have no universal applicability. But those patterns, those principles, those rules that are uniform and universal throughout existence, are they not at least the first derivative of the underlying noumenal absolute nature of existence?

Bloom extracts perhaps these very universal rules of how a seemingly Godless universe creates, but like Bohm[14] he cheats a bit: He looks at what is there, he captures the existing rule-pattern. Which requires a lot of intelligence, which is very smart, and act of highly developed consciousness, a very inventive synthesis indeed, but which nonetheless **is** not the designing of these rules. So where do these rules come from? An infinite regress?

No need for a bearded God and no need for six monkeys with six typewriters... BUT as far as we know it does require an intelligent screening/pruning algorithm to find the "causally invariant rules",

Stephen Wolfram is working on. It required the conscious attention of Stephen Wolfram himself to come up with sets of simple rules for his screening/pruning protocols. And guess what, what he finds are systems self-organised in nodes and cliques…how mind like, how fractal like… And how much like the condensation of nodes and cliques into matter described by Abraham and Roy[15] (Demystifying the Akasha) in their quantum vacuum theory, how much like Kaufman's[4] (Unified Reality Theory) and Campbell's[16] (My Big TOE) digital consciousness.

But there is more to the story of extracting the rules that Bloom has not touched upon. The whole business Bloom describes about the fact that it would be expected that a purely random universe after the big bang would yield particles in a zillion different shapes and sizes, but instead condenses to a parsimony of a mere 57 particles. A process repeated throughout the history of emergent patterns: The variety at every level (quarks, protons/neutrons/electrons, stars, galaxies, life forms) is rather limited.

A theory which may give a hint to a rationale for this limited number of hits in a screening protocol may be reflected in the principle of "Entropic attraction": the emergence of a limited number and forms of structures arising from the fusion of the elementary building blocks causes the maximal dissipation of energy. Yes, you read this well, Entropic energy dissipation is higher in the presence of a few structures than in more homogenous field of apparently more random elementary building blocks. Why? Because the aggregates create more space for dissipation for the remaining elements, because by clustering i.e. "attracting each other", the remaining elements can better dissipate their heat. And guess what, Verlinde[17] has shown that this form of entropic attraction can also be considered a computational process, which gives rise to… of all things Newtonian gravity.
So it would appear that Chaos drives the creation of levels of Order.
Really? Or do we need a paradigm shift, an "axiom flipping" as Bloom would call it, a Gestalt-switch.

Campbell[16] sees in the autopoietic morphogenesis of emergent properties, of qualities, what he and others would call negentropy. Not the negentropy of Shannon's information; that one lacks meaning. But the negentropy of meaningful information: the abstraction or extraction of essential qualities, which make one entity different from another. A diffeont: a set of ontological parameters to distinguish one pattern from another. Hey, that's what a Mind does when it tries to distinguish A from B. In my book TV1.0 I suggest an architecture for an artificial equivalent of consciousness for the internet, the essence of which is that websites are hierarchically embedded in Hubs of ever higher ontological categories, which Hubs extract which Websites get the most **attention**, (attention being a multidimensionally defined concept here).

Attention, as in attention given to successfully-pollen-collecting-bees in a beehive, attention as the abstraction of the essential, as the most successful heuristic, like the pillars of a termite hive. This extraction process of what is most essential to the existence of the Webmind is presented to higher ontologies, decision making routines of the Webmind, the upper ontologies of which could be considered as its (quasi)consciousness. Because it abstracts the stimuli and reacts thereon with a response. The higher the quality of this architecture, the more it optimises its interaction with its environment. And the structure is fractalised; lower ranking ontological Hubs still have capacity to carry out decisions unless overruled by a higher ranking Hub. Basically the structure reflects the potential of dealing with its environment. It is a type of a reservoir of highly ordered energy capable of highly ordered release thereof in its response.

**Syntropy**

In fact what I am arguing is that consciousness itself is a reservoir of highly ordered energy, huge amounts of energy with a very low entropy due to its structure, based on the attention it gets bottom-up and executing top-down in a highly structured way by focussing on what is most essential, what has the greatest optimised utility for the whole system. How it can reach a sustainable equilibrium with its environment for as long as possible. The inbuilt hunger for sustainable symbiosis as its emergent morality, because that is the most probable way for its survival.
And the higher the quality of the consciousness, the lower its entropy, resulting in more control when responding to the attention it receives in

the form of stimuli. And you may recognise this from your own ways of dealing with tasks: If you do it with love and care, with attention and control, the result is better and you enjoy the doing more. So if consciousness gives attention back, it autocatalytically improves itself by autopoietically generating joy. Giving attention results in receiving attention, from others and from your own mind.

Consciousness has therefore a strong relationship with pattern, with structure and it may well be that if we find the right formula to quantise consciousness quality as negentropy and weigh this against the heat dissipation created in a conscious creative process, that the ultimate outcome is that the heat dissipation entropy is more than compensated for by the gain in consciousness quality. And perhaps Weber Fechner's law is the right metaphor for both consciousness quality and pattern emergence. Perhaps the quality of consciousness is indeed reflected in the constant k in the Weber-Fechner formula[12] $S= k\ln A/A_0$ which should be summed over all the processes and senses involved. Because stimuli corresponding to greater differences can be dealt with by a consciousness with more versatility, with more resistance. Perhaps it is no coincidence that the formula of entropy, Boltzmann's $\Delta S=k \ln W/W_0$ is highly isomorphous to that of Weber-Fechner, as its symmetric counterpart in the world of matter. Perhaps the degree of randomness of an event is linked to the degree of order created in a process (without them necessarily being quantitatively equivalent).
Because opposites are joined at the hip.

And what do we see? The more structure, the more consciousness a species has, the more versatility it has, the more variety it can create, all aiding to maximise heat dissipation – and to build ever further structures until the system at a new meta-level becomes isomorphous again with a previous level far away: Until it becomes an abstractor per se: a nodal network of mindedness, and interconnected dendrogram.
And perhaps it is not entropy that causes the attraction resulting in levels of order, but its consciousness counterpart. Perhaps we mistake the rope for a snake. Perhaps the solution to the God problem is that there is no bearded God-as we-know-it, but that there is a fractal of consciousness, of self-repetition in ever increasing variety, ever increasing potential. An interference pattern of beautiful structures of

panprotopsychism, of psychic interacting social entities, particles and waves singing the harmonies of quantification, commune-icating. Which ultimately manifest at the highest semiotic level as a one transforming into a zero and back again ad infinitum, an ever repeating bagel and stick of Bloom's toroidal model of the universe… part of a yet bigger matrix of digitality, which is only part of the fractal the Maya-fabric of consciousness uses for self-expression. Self-expression of the simple alchemical rule “solve et coagula”, differentiate and integrate. How to differentiate optimally? By using meta-transformers transforming transformers into new ones following the optimal asymmetry ratio parameter Φ (1.618) of beauty. The same ratio parameter that forms the basis of Fibonacci's spirals. Spirals which are the semiotic expression of logarithmicity and parsimony. The logos of the arithmos: The ultimate ratio expressed in number. How to integrate? By attracting harmonies, clustering attention of admiration of inequalities. By screening for the highest pile being built and focussing your attention there and pruning the not so successful recruitment strategies away, by recycling them. By creating meaning in the form of Bayesian proximity co-occurrence, a Yoneda didensity love relationship shaping a new ontology.

The purpose of this level of the game Leela ( in Hinduism it is believed that existence is the game or play of the supreme consciousness, which game is called "Leela") is to find out the rules of the game. The purpose of the next level of the game Leela (self-sustention and self-generation: autopoiesis) is to find the purpose of the game (resulting in an infinite loop of self-redefinition).

Therefore, Bloom's materialistic stance on how a Godless universe intelligently creates in fact is based on a hidden assumption that primordial consciousness is the ultimate ground of existence.

## Chapter 4 The Substance of Emptiness

An attempt to reconcile notions of Buddhism and Hinduism.

A while ago I went to a lecture of the Dalai Lama, where I encountered some concepts that seemed to be in contradiction with the teachings from Advaita Vedanta (non-dual branch of Hinduism: we are One but not the same), which I consider to be my most promising springboard.

### Background

Firstly, according to Buddhism no phenomenon has an ultimate substance. Nagarjuna claimed that such phenomena are empty and considered all experienced phenomena as "dependently arisen" from the emptiness called "Shunyata". (Are you not thinking, what I am not thinking?). Yet he did not mean that such phenomena cannot be experienced or that they would be non-existent. He rather meant that they are devoid of an eternal, permanent substance (Svabhava). In that way it is said that Buddhism should not be confused with nihilism. In Hinduism the ultimate "substance", which is not really a substance, is primordial consciousness itself, Purusha.

Secondly, in Buddhism there is ultimately no "Self" ("Anatta" or "Anatman"), which would be the underlying aspect of being, the "hypostasis". This is strongly contrasted with Hinduism, in which the "Self" is deemed the ultimate ground of being, the all-pervasive omnipresent Brahman (God) or primordial consciousness from which all phenomena arise.

Thirdly, he said that "in Buddhism there is no creator". In different Hindu sects there are different mythological descriptions of the creation process, and different names for a creator, but the role of higher intelligence in this process is undeniable.

Let's see if a comparison with modern notions from quantum physics and ontology would be able to help us to reconcile these seemingly opposite stances.

**Is there an ultimate Substance?**

In quantum physics there is the wave-particle duality. In this duality, considered in the light of Einstein's equation $E=mc^2$, energy (E) can either be unbound i.e. in a wave form (like electromagnetic radiation) or material in the form of e.g. a subatomic particle, such as an electron having mass (m). Energy in its particle form can only be detected by an appropriate instrument configuration, but if a different instrument configuration is used, a wave-like behaviour is observed. What seems to be solid and material is in fact a buzzing beehive of energy streams that form orbitals, shape and structure together.

What is important in this last phrase is the word "together". As long as energy is alone and not observed it is presumed to be non-local, in other words everywhere. Upon measurement, observation, the wave collapses to make a particle observable. This requires and interaction of that energy with a material configuration of a sensor. Together the observer and the energy to be observed are capable to manifest an observable local, particulate entity. This nicely fits the notion of dependent arising for the observable energy.

It should be realised that all type of matter is a collection, a congregation of multiple particles or energy packages. There is no such thing as an unobserved free particle. Rather, as long as it is not observed, it is non-local. Materiality therefore requires at least two energetic entities that together establish a kind of density, an interference pattern, which is locally concentrated.

There must be an observation for the particle aspect to be observable. Yet if you leave a two-slit experiment to take place in your absence, the experiment still takes place and you can still see the outcome, even if you were not present at the moment of interaction between the detector and the energy wave collapse that established the particle. This means that the terminology "observation" does not necessarily imply that the observer is human. Rather, the interaction between structural matter and functional energy transmission coming from the detector on the one hand and the energy to be observed on the other hand, can be considered as an observation.

In view of this interaction aspect of the observation, one could state that matter is the consequence of the mutual observation of energetic entities; the interaction of at least two energetic entities forming a so-called "didensity".

Interestingly enough, in the world of information a given entity, a thing can only ontologically be defined by at least two descriptive statements. Meaning is conveyed only by a didensity of informative content. A term without a relation to another term is just a name, with no inherent essence. It is only when things are defined in terms of the elements that constitute them, that they can be understood as a thing by the brain. Yet we also know certain sensory qualities (such as colours, tastes, sounds) which a priori might seem to escape from this dual type definition. This is not so. We can only know a colour if we know at least another colour, we can only get some information out of sounds if they contrasted to other sounds. If there is no contrast there is perhaps observation, but there is no meaning. Meaning can only arise if there is contrast. A didensity. A togetherness of energetic entities, which at least differ in one aspect (e.g. their relative location) of a quality (the aspect), which is the same for both (this sameness of quality which only differs in degree is called the "identity of opposites" by SpinBitz[18]). Thus a quality is polarised into a duality to generate a phenomenon. This neatly fits the notion of "dependent arising".

Yet energy, which is not materially bounded, can still convey information, if this energy transmits a contrast. An interference pattern, a pulse sequence, a set of different frequencies. The most stunning modern application perhaps being the Wi-Fi, which transmits very complex information non-materially. This transmitted apparently unbound energy definitely has a structure. And if it has a structure, it also has a certain pattern, shape or form. So even in the apparent non-material world, which can still be decoded into materially observables, there must be some kind of structure and form. This implies that it is only relatively non-local. It is perhaps non-local with regard to the scale of our instruments and ourselves, but from a cosmic view point, if you were able to see, be or feel that energy, it would still be a wave form with a shape advancing or expanding through space. It would for such an observer be "particulate", limited in time and space. The peaks and valleys of the interference pattern would be "somewhere".

In a sense, such energy would be "material", a certain substance, as it apparently has implicit structure and form. So materiality is perhaps but a relative term. Indeed, what we consider as solid matter is mostly empty and the subatomic particles that are left are just whirlwinds of energy revolutions at light speed in a –for us- very limited space. So matter itself is in fact nothing but energy. We can accelerate the subatomic particles until they disintegrate into pure energy, which we then measure in Mega electronvolt units. For an observer, who is billion times smaller than a subatomic particle, such energy whirlwinds that build the particle would be non-local unbound energy.

So it seems as if matter and energy are relative terms and that there is no end to this tower of turtles of emptiness being form being emptiness etc. depending on the scale that you look at it.

So it is then equally valid to state that form is as substantive as emptiness. One can say ultimately there is no emptiness; if you were able to look further down, e.g. below Planck scale, there would be an infinity of levels of materiality. Likewise one can state ultimately there is no materiality, if you were able to look further down, e.g. below Planck scale, there would be an infinity of levels of unbound energy, which is nothing else than emptiness.

So it seems "empty" is not so empty after all, and "form" is not so full or permanent after all. Even the polar notions of form and emptiness seem to follow that pattern of "dependent arising".

On the other hand these arguments are speculative. Perhaps there is a lowest level of aggregation below which there is only unbound energy in revolutions. For us it would seem that that level is the Planck scale, but we can't be sure about that.

The point I wanted to make is that apparent unbound energy (i.e. energy not bound in a material form) has structure, has an interference pattern and does convey some information.

Modern analyses of the vacuum have shown that it is not so empty. The well-known Casimir effect shows that particles can arise from a vacuum. Which then leads to the conclusion that the vacuum is a kind of energy sea boiling with activity. This is sometimes also referred to as the zero-point energy.

The physicist Nassim Haramein[19] suggests that the vacuum does have a structure, namely that of an isotropic vector matrix, which can be best modelled in the form of cuboctahedrons, which themselves can be composed of adjacent octahedrons and tetrahedrons. Or in a dynamic way as a "Jitterbugging" octahedron-cuboctahedron, which when closed is an octahedron and when fully open is a cuboctahedron. This non-stop "Jitterbugging" would generate a sustained toroidal flux, which establishes the subquantum zero-point field.

Although these theories have not yet been proven, they are a very elegant approach to describe the possible "structure" of emptiness. It is in a certain way the return of the "ether" that was denied by the scientists at the beginning of this century. Noteworthy, Nikola Tesla, who is the father of all wireless energy transmissions, was convinced of the existence of an "ether" and Einstein later started to doubt his earlier denial of the ether.

Interestingly, Nassim Haramein mentions the possibility of a fractal type nesting of octahedron-cuboctahedron, giving the vacuum an infinite structure; It is turtles all the way down. Like in the Hindu parable that the world rests on a turtle (or an elephant) and when asked what this turtle rests on the answer is: another turtle. When the question is repeated the answer is: It is turtles all the way down.
So the answer to the question "is there an ultimate substance" can perhaps be replied in the form of an infinite regress zero-point energy matrix à la Haramein, with alternating form and emptiness. As that system is in a constant flux of expanding and retracting (Jitterbugging), and perhaps even spinning, there is no place which has complete emptiness of energy flux, nor is there ultimate permanency of one form or structure. Rather the forms/structures alternate and recur at intervals, rendering them "semi-permanent" or "dynamic" if you wish.

So if this turns out to be the ultimate truth, Nagarjuna was neither right nor wrong or both right and wrong simultaneously, as regards the Shunyata.

I will come back to this issue after having discussed the next issue, which provides some further clues as regards the notion of an ultimate substance.

**Is there a Self?**

As regards the Anatman (or anatta)-atman (or Brahman) dichotomy, I think that Buddha wanted to fight the "false ego". The idea that our personality can be kept permanently. The idea that we have an individual soul independent from the Brahman.

To resolve this issue I have to go deeper in this matter. Forgive me if I have to repeat some of the concepts mentioned above in different words.

With the logic analytical technique of "Syndiffeonesis", developed Chris Langan[3], it can be demonstrated that everything is reductively the same.

"Syndiffeonic" means "difference in sameness". Any assertion to the effect that two things are different implies that they are reductively the same. The difference between two things can be described in terms of quantities and qualities of something they have in common. This difference map builds the relations. If you do this for all things, and if you do it recursively as regards the differences between the differences of two or more sets of relations, it turns out that all things are reductively the same.

The difference can be said to be written in a common language (not necessarily a literal language, but proverbially spoken) of a quality that is the same for everything. The mere fact that a difference between things can be described linguistically (or geometrically, which is just another type of language) implies that difference is only "partial" and quantifiable in terms of "contribution. Both related things ("relands") are quantified manifestations of one and the same quality.

Which means that relations lead to patterns, patterns of form, forming some kind of information. Unlike the Greeks who found atoms to be the ultimate building blocks which cannot be reduced any further, today our paradigm is that pure energetic (structured) vibrations (as described above) appear to be the ultimate essence of reality, which cannot be reduced any further. As things are only temporarily compounded of one and the same quality but eventually return to that same original energetic state, things can be said to be "in formation".

When things interact they influence each other, leading to a mutual "reaction". In order to be able to react, some information must be exchanged. Between two colliding billiard balls a vector impulse is transmitted and exchanged. Electric charges "feel" each other's presence and react correspondingly in their movement; again some kind of energetic vibration is transmitted. Objects, albeit in a very rudimentary form, are somehow "aware" of what is happening to them when they encounter other objects and "react" thereto. There is some kind of "sensing" involved. It is true that this is not the focussed aware-type of sensing that living entities (as we know them) can have, but it can be called sensing, perceiving in a certain way. Even if it is programmed into the object how to react, when encountering a certain stimulus, the fact that it can react to a stimulus, means that it must somehow be able to perceive that stimulus. We can call this ability to perceive a form of proto-awareness. As all forms of energy in one way or another interact with each other, even if it is in a very minute almost imperceptible way, we cannot deny that there is an interaction; there is a form of sensing involved.

In other words the pure energetic structured vibrations, which are the ultimate essence of reality and which cannot be reduced any further, have an intrinsic quality of proto-awareness. They are a presence, which senses, a "presense" if you wish, for as long as they have not encountered any other presence, they may not sense anything.

But as long as pure energy has not encountered any other energy, it is in its wave state and isn't really localised anywhere. It is only when energy interacts with something else or another form of energy, that it becomes manifest and localised. Without interaction, not much can be said about energy. When it interacts, it means that there is at least a second entity to interact with. A meaningful event in the life of an energy beam can only occur when it interacts with something else, when there is a proximity between the two; they can be said to co-occur or to coincide if they are sufficiently proximate to be able to influence each other.

Funny enough in modern programming building towards a "thinking" robot or computer network, such as Watson from IBM (this program searches answers an looks for "meaning" via so-called "Latent Semantic Analysis"), "meaning" is derived when two terms have a statistical "proximity co-occurrence" in a body of text.

Similar to the building an ontology, both in the interaction of energies or (sub)atomic particles and in Latent Semantic analysis, we encounter the same thing: Meaning, a meaningful interaction, a reaction can only occur if there is sufficient proximity, to localise an event, an occurrence, something that gives meaning.
The undifferentiated energy as such appears a priori meaningless and it can only form an event, something perceptible once it encounters another energetic vibration of some sort. So any event is minimally a di-density, a proximity co-occurrence. Only the relation is observed in fact. The thing as such remains unknowable for the moment; Kant's Noumenon.

Wittgenstein and other philosophers pushed this idea even further: the only facts that can be said to exist are the relations between the things; the things as such have no "independent reality". If you consider a three-dimensional sea of energy, as long as it is homogeneous everywhere, no things can be said to be observable. But when you have interference patterns in the energetic sea, it means that there has been some stimulus to form an inhomogeneous distribution in the sea. The different energy waves can then interact, giving rise to observable interference patterns.

What we can observe, we say it exists; it stands out from a more or less homogeneous background. But that does not mean that that background is an absolute void, an absolute nothingness. Rather it is bursting with potential energy waiting for its chance to interact. Haramein's isotropic vector matrix.
So observable "things" can be said to be the consequence of the relation between (at least) two streams of energy. Only their relations then lead to observable existences, the ground of which is not directly knowable via sensory perception, although it can be inferred if you depart from the millennia old concept "nothing can come from nothing".

So in its most fundamental ground form existence originates from the ability of energetic vibrations to sense stimuli (proto-awareness) and to interact/react so as to give rise to interference patterns: to shape and to form, which generates a relation event, which is observable, which we can call a form of information. So energy is endowed with proto-awareness and proto-information (i.e. the ability to sense and the ability to interact and thereby give form and shape). Thus, we might have a clue here that the most fundamental ground of being is not substantial in the sense of matter or structured energy, but not an absolute void either. Rather it may be proto-awareness or primordial consciousness per se.

Everything that can be said to "ex-sist" (i.e. "stand out" from a background as opposed to "subsist": being the underlying background, which is not an absolute "nothing") can therefore be said to be a compound of at least two different streams of energy. Everything that can be said to "ex-sist" can be said to be of a temporary nature as it ultimately dissolves back into its structured energetic building blocks. But its underlying energetic proto-awareness and proto-informative ability is never lost.

Proto-awareness leads to self-creation, self-organisation self-sustention (autopoiesis), which is a cybernetic stimulus-response-feedback loop inherent to Reality. Reality has sensors, senses what happens otherwise it is not possible to evaluate conditions, recognise these and respond thereto. This appears to be true at any level: wave-energetic, subatomic,

atomic, molecular, macromolecular, cellular, organ level, plant and animal level.

So awareness of some sort, i.e. consciousness, however minute, is an inherent functional characteristic of everything that is. This leads us to the need to accept the notion of hylozoism or panpsychism, wherein every energetic entity is inherently endowed with a form of (proto)-awareness. For if it did not or could not interact with other energies/entities, it would not exist.

As long as energies/entities are localised and autopoietically work for their self-sustention, they can be said to be "selfish". But as soon as they start to contribute to the creation, organisation and sustention of other entities, they start to merge with existence and the "self" aspect starts to diminish. In that sense we have ultimately no "individual self", an eternal individual atman, which remains limited to one specific form, as our energies will one day merge and meld and contribute to a greater whole together with other energies. The drop of energy, that we temporarily are, will one day flow back to ocean where it came from.

But that ocean as a whole is likely to be aware of its internal energy streams. That ocean as a whole is a flux of consciousness at a higher level. That ocean can then be equated with Brahman or God if you wish, being all-pervading. And that is what the Upanishads (a set of Hindu scriptures) call the (higher) "Self". But as it is the infinite whole of all existence and subsistence, there is no point in calling this "self", because "self" would imply the existence of "non-self". And there cannot be anything outside the whole of existence/reality. Reality is that which contains all and only that which is real. There are no things or beings outside of reality for if they are real, they are per definition included in reality. Whatever can influence reality is per definition part of reality. This excludes external causes. But this also means that the term "self" for Brahman is in fact pointless.

So perhaps Buddha did not deny the existence of Brahman, but realised that the term "self" was inappropriate for the whole and that the individual atman was no permanent entity. Noteworthy in Hinduism Buddha is revered as one of the incarnations of Vishnu.

Interestingly enough the term "atman" is etymologically related to the German "atmen" which means to breathe, and in fact the "Jitterbugging" structure of the vacuum can be considered as a form of breathing. As all phenomena arise and subside endlessly from this sea of energy, this can also be considered as a form of "breathing".

In that way perhaps there is no real dichotomy between these aspects of Buddhism and Advaita Vedanta (non-dual Hinduism), but only an apparent one, as the same words are used, but slightly different meanings intended.

Noteworthy, it is my hypothesis that the ultimate ground for existence (which itself by definition does not "exist" (because it cannot be differentiated from anything else as it is absolute) but rather "subsists" is the feedback-loop which makes that energies can form relatively stable informational structures, like the strings or branes in string theory, or even a toroidal shape, which nest into themselves. It is this self-sustaining feedback-loop, which I hypothesise IS consciousness. The feedback-loop giving rise -to what Giulio Tononi- calls "integrated information" as essence of consciousness. Energies can only be observable, can only exist, if they can sustain themselves by self-resonance, by impinging on themselves, by forming a standing wave. Self-interfering, self-referring, self-impinging, self-integrating, self-resonating: all synonyms of the same process. It is this impinging, this contact with itself which is experienced as self-awareness. Energy and Consciousness are then two sides of the same coin, which I would like to baptise "**Conscienergy**".

And it is this impinging on itself by energies, which also creates a certain density of energy, a peak in a wave in an interference pattern, a semi-localised substance or materiality if you wish. In other words energy and consciousness are perhaps nothing but aspects of the same all-pervading essence of being, nameless and formless, yet builder of all forms and names. What we call matter is just a particularly dense form of thereof. A "Self" of Soul/Atman may not be more than the illusion of a local string of conscienergy that it would not be connected to the rest of Conscienergy, the "Whole" or the "Absolute". In that sense there would be an "end" to the apparent fractal of matter and free energy. Maybe Conscienergy even has a foundational resting state form, which according to Ithzak Bentov would be the "Torus", which then would fit the Hindu notion of the "Svarupa" as the true or essential form of Consciousness itself.

Oh, and by the way, don't try to "understand" the Whole or the Absolute via the intellect. You will fail. Understanding occurs per definition in terms of constituent parts, at least three of them according to Buckmister fuller. So if Conscienergy is the foundation of the absolute whole, it has no "constituent" parts and cannot be intellectually understood. Logic was designed to reason in terms of specific parts or instances from which a generalised pattern can be abstracted. But if the Absolute or foundation IS the multiversal pattern itself, you cannot use it as a specific "part" or "instance" to build your reasoning upon. Per definition.

**Is there a Creator?**

The third topic relates to the notion of a “creator”. In the more mythological branch of Hinduism as described in the Vaishnava Puranas, the demi-God Brahma (not to be confused with Brahman) created the world or universe as we know it. Yet Brahma himself originated from the navel of Vishnu, an origin he couldn’t find himself. When the incarnation Krishna of Vishnu confronted Brahma with countless other Brahma’s from parallel universes he was flabbergasted. Shiva was born from the head of Brahma in the Puranas. This is the view of those who see Vishnu as the highest God and equal to the highest Brahman, the view of the Vaishnavas. The Shivaites on the other hand describe how Brahma and Vishnu were seeking the origin of the Lingam of Shiva but could not find it. Although the different sects may be divided over the name of the highest God, Hinduism accepts that all is One and all is a creation of one ultimate entity, commonly denoted as the Brahman.
According to the Dalai Lama, “in Buddhism there is no creator” and existence is the consequence of the “dependent arising”.

The statement that in “Buddhism there is no creator” is however not the same as “there is no creator” or “Buddha denied the existence of a creator". It seems that for the philosophy of Buddhism, the notion of a creator is not a necessary concept.

In fact Buddhism was in a certain way a reaction to an ancient form of Hinduism also called Brahmanism, in which there were frequent animal sacrifices to please the Gods. Buddhism may have been a necessary reaction to stop this cruelty, which also seems strongly contrary to the morality of the more modern abstract form of Hinduism i.e. Advaita Vedanta. In Advaita Vedanta the most important key concept of morality is "ahimsa" or the absence of violence. This includes refraining from killing animals and that's why most Hindus are vegetarian.

Anyway, the all too frequent animal sacrifices in early Hinduism were a kind of perversion and were not in line with the concept of all-is-one. It was therefore important to dethrone the Hindu Gods from their pedestal and it may well be that that is the reason why Buddha never addressed the topic of a God or Creator.

If we observe however the complex structure of our solar system, which is seeded with a great number of kinds of numerical clues hidden in the measures of the orbits and circumferences of the planets, it is difficult to deny the existence of a higher intelligence, who must have designed this. John Martineau[20] has described a great number of coincidences in the solar system, which are so astonishingly precise that they defy the notion of spontaneous arising. It is already quite a coincidence that the Moon has a distance from the Earth and the Sun and a size such that it can exactly cover the Sun when seen from the Earth during an eclipse. This is often dismissed as a form of the "anthropic principle", which states that it is "unremarkable that the universe's fundamental constants happen to fall within the narrow range thought to be compatible with life". Things seem so incredibly coincidental, for if they were not we wouldn't be here to observe it.

Well, what if I tell you that Mercury and Earth's mean orbits are in exactly the same relation, ratio as their physical sizes and what if I tell you that the same is true for the Earth vs. Saturn. That an octagram and a fifteen-pointed star can be drawn respectively in these respective sets of orbits/circumferences, wherein the points of the star precisely touch the orbit or circumference of the greater planet and wherein the inner space of the star precisely touches the orbit or circumference of the smaller planet, you may start to frown. If I tell you that this star also produces the exact tilt of the Earth, you may start to wonder what is going on here. This is just the beginning. If you study the complete solar system, you stumble on numerous of such coincidental relations, which on top of it, describe the most beautiful flower like patterns.

And this is not the end of your astonishment: Nature is full of bizarre coincidences, and our measures like mile, kilometre feet and centimetre seem to have been magically chosen to encode our base ten numerical system. And the ancients appear to have been aware of this and have encoded this in their buildings.

The equation[21] $\{(\text{Hlf} \times \pi)/\Omega\} = \{c\}$ ({} represent numerical values), describes a numerical relationship between the numerical value of hydrogen fine transition line, the ratio between the circumference and diameter of a circle, and the numerical value of the speed of light in a vacuum in Thoms/sec, wherein omega is 0.012345679012345679 (by multiplying the 0.012345679012345679 by the missing 8, we get 0.0987654321 –ergo omega encodes base 10 number system).

The speed of light is encoded three times in the pyramid of Gizeh. The Great pyramid: GP, has 144000 casing stones (144000 is the speed of light in Earth grid arcs/grid second); The position in degrees latitude of the complex of pyramids halfway Khufu and Khafre is 29,9792458, which contains the same number sequence as the numerical value of the speed of light in meters per second; the difference between the outer and inner circumference of GP also encodes the light speed (299.8 m). The height of the GP (280 royal cubits: Pi - $\text{Phi}^2$ = royal cubit) encodes the distance between Earth and sun and also the polar radius of the Earth.

The base length of GP is 365.2422 sacred cubits, which refers to length of year. Other sizes used in the GP encode moreover the radius of Moon and Earth, Pi, Phi etc. The diameters of the Earth and Moon (7920 miles (7920=11x6!) and 2160 miles (2160=3x6!), respectively)

are in the ratio of 11 to 3, which proportions also encode the proportions of the human body (as in Da Vinci's Vitruvian man). Twice the perimeter of the bottom of the granite coffer times $10^8$ is the Sun's mean radius.
The radius of Earth and Moon together equal 5040 miles which is 7!, but also 7x8x9x10. Ergo these radii together encode base 10 number system as well. The circumference of the Moon is $12^7$ feet. (Chapter 2 of part 2 will go deeper into this subject and show you even more coincidences).

And I can go on and on. How could the ancients have been aware of the metric an mileage system? How could they have known the speed of light? How can it be that constants and planetary sizes encode our base ten system?

Do you realise how extremely fine-tuned the materials chosen to create the proportional coincidences of Mercury, Earth and Saturn must be? It defies understanding.

To me it seems all too coincidental and it looks more like a very elaborate beautiful work of higher intelligence, which expresses itself *inter alia* via us. We may not have known what we were doing when we built these structures, but a higher intelligence with knowledge about the future seems to have been working through us.

As if the system is screaming to be discovered, to make itself known to us. It appears very likely that there is a higher intelligence and that this higher intelligence is making all creations. Primordial consciousness appears to be expressing itself to become known by its lower temporary "selves".

Whether this is the work of one higher intelligence or multiple collaborating intelligences is not so important. It just seems to be too farfetched to assume that all these measures and ratios are the consequence of "spontaneous arising". Therefore to claim that there is no creator at all is in view of these overwhelming data a less likely hypothesis than the opposite.

**Non-Dualism**

It is noteworthy that Nagarjuna never said that "dependent arising" and "emptiness" represented the absolute truth. He used these notions pragmatically and used a higher form of logic to show that via logic, reason, metaphysics and science we cannot know the absolute truth. Note that he did not say that we cannot know the absolute truth. In many religious traditions the absolute truth is said to be knowable only through a direct mystical communion. Samadhi in Hinduism, Satori in Zen-Buddhism.

Alan Wallace[22], a Buddhist repeatedly writes that consciousness is a relative thing. But at a certain point, he makes the distinction between "substrate consciousness" (relative consciousness; awareness of a phenomenon) and "primordial consciousness". He quotes the Tibetan monk Düdjom Lingpa: "Primordial consciousness is self-originating, naturally clear, free of outer and inner obscuration; it is the all-pervasive, radiant, clear infinity of space, free of contamination".

That notion perfectly matches Advaita Vedanta. It is this notion that Hinduism, perhaps unduly, calls the "Self" or Brahman. So if you dig deep enough, beyond the veil of semantic obscuration, it turns out Buddhism and Hinduism aren't so different after all. And the most promising thing is that according to Vedanta : You are One with that! Tat Tvam Asi. You can experience merging and becoming one with that consciousness.

It is not a miracle by the way, that notions from Buddhism can be reconciled with Advaita Vedanta since they are both non-dualistic paths. Moreover, Shankaracharya, the author of the Advaita vedanta philosophy hoped to reintegrate Buddhism into Hinduism via Advaita Vedanta.

If you're at the highest level of consciousness, clap one hand.

## Chapter 5 Metaphilosophy

An essay on the (f)utility of philosophising.

The School of Athens by Raphael, 1511.

My friends sometimes tell me that I shouldn't philosophise too much but rather enjoy and experience life directly in a state of mindlessness. I have counter argued that such an attitude is only possible once you have convinced yourself of the futility of philosophising, which apparently is a process that you need to go through via the very medium of philosophy, which is reason.

The purpose of this chapter is to explore for myself the (f)utility of philosophising as a means to come to "correct knowledge", which Patanjali[23] calls "Pramana" in the Yoga Sutras by reasoning this out in a quasi-philosophical manner. I choose not to follow the traditional methodology of philosophy for reasons that will become clear in the course of this chapter. Although I ultimately desire to develop my own alternative methodology, the present chapter is a first exploratory attempt. A first brainstorm to order my thoughts, which by no means I claim to be exhaustive.

Whenever we use the word "philosophising" we have a certain meaning for this word in mind. Although each individual probably has his/her own definition of this terminology, for the sake of this chapter I distinguish two classes of philosophising:

1) Philosophising by layman, which essentially amounts to reasoning and arguing about certain mental concepts, based on ill or fuzzy defined definitions and which relies on a non-systematic way of reasoning, which is allegedly based on "common-sense".

2) Academic philosophy. As to this form of philosophy, Wikipedia gives a definition: "Philosophy is the study of general and fundamental problems, such as those connected with reality, existence, knowledge, values, reason, mind, and language. Philosophy is distinguished from other ways of addressing such problems by its critical, generally systematic approach and its reliance on rational argument."

I did not study philosophy, so my type of philosophising appears *a priori* to fall in the first category. But I hope to be able to show by rational arguments based on my common sense, that both methods have their inherent flaws or at least are ultimately futile in their attempts to come to "correct knowledge", in the sense that Patanjali uses the word in the Yoga Sutras. An analysis of Pramana, will have to wait until the end of this chapter however.

As such this attempt is a kind of "philosophising about philosophy", which makes it a kind of Meta-philosophy. Wikipedia defines this as follows: **Metaphilosophy** (sometimes called **philosophy of philosophy)** is 'the investigation of the nature of philosophy.' Its subject matter includes the aims of philosophy, the boundaries of philosophy, and its methods. It is considered by some to be a subject apart from philosophy, while others see it as automatically a part of philosophy.

In this sense my present meta-philosophical attempt is not futile, that -if it works out well– will save me from wasting time on futile future philosophising and possibly make clear which type of philosophising has utility for me. In this sense it is not part of academic philosophy, in that I intentionally choose to avoid the "generally systematic approach" of academic philosophy, whilst still relying on the rational argument.

One of the problems with the academic approach (as the ruling thesis of what philosophy is supposed to be) is that an essential part of its general systematic approach relies on providing new definitions of the terminologies used.

Although it is necessary to clearly know what one is talking about, academic philosophy often loses itself in a typical association-type think fever, the quagmire of semantics, leading to hopelessly long lists of definitions, before you have even started to reason. Although cumbersome, time-consuming and rendering the text to be read utterly boring, it seems an indispensable pre-condition.

But it often leads away from the very concept that one wants to study. Because every definition becomes a topic of philosophical study itself before one can get to the very concept that one wants to discuss. A kind of runaway of philosophical spin-offs of all the parts that are needed to describe a whole. This can lead to chicken-egg problems when definition of concepts are interdependent; where you need the chicken to define the egg and the egg to define the chicken, so that in the end you do not have a meaningful delimitation of either concept (and you can only merge the concepts into an interdependent meta-concept).

Because every terminology is described in terms of other terminologies, you get a repeating process where you probably can't stop until you have given philosophical definitions of all the words in the dictionary. As academic philosophy is incomplete as regards this, it fails to properly apply its own methodology and is bound to work with common-sense and intuitive meanings of terminologies, sometimes without even being aware of that.

But there is a worse problem here: namely that the meanings of the very terminologies you wanted to use to describe a concept have been so distorted due to the academic defining process, that they are no longer suitable to define/describe/analyse that concept.

What we often see, is that the accepted philosophical meaning of a terminology (i.e. accepted by the ruling paradigm in academic philosophy) is very far away from the instinctive or common sense meaning of that terminology. Whereas the original aim may have been to clarify an instinctive or common sense concept, the final concept with the same name that academic philosophy is describing, is no longer identical to the topic that one wanted to treat. A serendipitously generated self-consistent piece of philosophy may have been generated, but the concept they deal with, the concepts they have defined, do not reflect well the instinctive or common sense meaning of that terminology. What Heidegger[24] understands about "being", "beyng", "Dasein", "Mitsein", "Existenz" etc. has very little in common, with what you or I instinctively sense as the meaning of "being" and "existence". The funny thing is that the academic philosophers are in a sense aware of these distortions, so that they use brackets, diacritical marks, and other symbols or slightly change the spelling of the terms like "beyng" (Heidegger) or "differ**a**nce" instead of "difference" (Derrida).

Philosophers then have to go through a cumbersome process of discussing all different types of definitions given by different philosophers to a terminology, which terminology is for them the best approach of "instinctive concept" that they want to study, to finally try to give it their own subjective meaning. And I hope they do arrive at a final meaning at all, because I get the impression, that much academic philosophy misses this point: The philosophical process transforms the meanings of the concepts so much, that they no longer correspond to the original concept one wanted to ponder.

This shows that even academic philosophy is a highly subjective process. The meaning of terminologies is changing over time as the ruling paradigms change over time. Then there are attitudes of showing-off how smart and how complex one can reason. And it certainly doesn't help to clarify things. You can only read academic philosophy texts if you're a philosopher yourself, they are hopelessly complex and do not well describe the point they want to make. I certainly don't feel attracted to this obligation of having to go through everything that has been said in the literature on a given concept before

I can make up my own mind on it. I'll even put it in stronger terms: This process stifles your ways of getting a clear understanding of a concept.
(No I don't want to define "concept" at this moment).

Perhaps I can illustrate what I mean with the following: I had studied classical guitar for many years, when I wanted to learn how to improvise. In the beginning this was not an easy process, because I was biased by all the melodic and rhythmic fragments that I had automatised in my study. I had developed a kind of blind spot for the possibility of new combinations. A friend of me, who had just started playing guitar, was composing the most interesting melodies and rhythms in jazz and blues and largely outperformed me when it came to improvising in this style. I had to "learn" "a vocabulary" of "melodic phrases" (licks) in jazz and blues in order to be able to jam with him. But it took a very long time before I started to develop my own set of licks and before I was able to spontaneously improvise new licks in the process of playing, based on hearing and feeling. I had the disadvantage of the so-called head-start. And in a certain way, for every skill such a disadvantage of a "head-start" can be present, including in philosophy.

Laymen philosophy (as antithesis) as I already said, suffers from ill or fuzzy defined definitions and relies on a non-systematic way of reasoning. Every amateur philosopher has his/her own instinctive definitions, which he/she has not clearly defined in terminological framework. This makes it very difficult to communicate. As everybody has had a different education and a different life-experience, the instinctive meanings of words given by different persons do not match. This is the basic source of almost all miscommunication in the world: The false assumption that our personal and cultural dictionaries match.

You can try starting your own philosophical inquiry into the nature of your experience, but as long as -at least for yourself- you have not clearly defined for yourself what you mean by the terminologies you use in your reasoning, you are bound to end up with fallacies.

If you have the rigour of going through the definition process to build your personal philosophical dictionary and vocabulary and you are careful not to diverge from your instinctive concepts by the seductive flow of redefining terminologies in ways that they no longer correspond to your initial instinctive process, you end up with a pair of extremely subjective philosophical glasses. They may help you to understand yourself, but they are worthless to share in communication, because the set of definitions is probably so boring, that no one will ever take the effort to read them.

But at least you may have gained some insight in your usual fallacies, so that you can avoid them.

Then there is still the danger that your way of reasoning is not following the principles of what is academically understood as reasoning and that you are introducing fallacies in you line of reasoning because you are not even aware of these fallacies. Here is a list of commonly known fallacies: https://en.wikipedia.org/wiki/List_of_fallacies, hereby incorporated by reference in its entirety. I will not be going too deeply into this topic. It is well-known that logic, the basis of reasoning has its own limitations. But at least there is a subset of logic, which when applied in a correct way, gives reliable results in the majority of cases. At least this part of academic philosophy is important to study and to incorporate as a mastered vocabulary. It is a pre-condition for any attempt to philosophise.

Anyway, it is not because there are parts of the philosophical process that are inherently sound, that therewith the whole become sound and non-futile. For the whole to be sound, all the parts must be sound. In other words, it suffices to undermine one part of the academic philosophical methodology in order to render it useless.

An issue with reasoning that is more problematic than the "logical process", is the fact that (both in academic and layman philosophy) the premises of the logical argument themselves have not always been proven to be sound, correct or true.

The layman is often even not aware that the knowledge about the premises used is incomplete and that therefore the premise is not necessarily true. Worse, certain premises not only have not been verified, sometimes by their very nature they are unverifiable.

This problem reaches its culmination in "speculative premises", which are hopeless starting points to build a solid argument. This leads us to further difficult philosophical issues of what is "truth", what is "proof" etc., which I do not want to define here.

The academic philosopher can avoid such issues by first going through the whole process of philosophy for each of these terminologies, ending up with definitions, which are perhaps internally consistent, but which do no longer "feel" like being representative of "truth" or "proof".

But in science as an extension of philosophy many premises are speculative. The very concept of a hypothesis is based on speculating what might happen. Therefore the scientific method uses a methodology to prove a hypothesis.

One of the worst problems with science as an extension of philosophy is that it has never proven its most basic tenet: That something must be proven by the scientific method for it to be true. It is rather so by definition. But that is a kind of logical fallacy as well: To do away with a problem, by making the problem part of the definition.

I do not wish to enter the discussion of what "truth", "proof", "being", "absolute", "relative", "reality", "illusion" etc. mean, because that is part of a philosophy itself. The purpose of this essay is to shed light on the futility of philosophising as such.

One thing I do wish to say about science, is that it is largely "inductive": It suggests a pattern based on "strong evidence", which gives it a certain probability. You get a cloud of data-dots and you connect the dots in a certain way, from which you derive an abstraction, a general trend or a certain correlation of two observable parameters. But the way you connect the dots heavily depends on your hypothesis. Aliasing shows, that there is often more than one way to connect the dots, and unless you are aware of that, you may be tempted to draw a straight line through every cloud, where perhaps a polynomial, a hyperbole, a sine or another mathematical expression form would have been more reflective of the underlying reality. There is only one logical rational process that gives irrefutable outcomes and that is deduction. Induction can at best predict a probable outcome.

In fact the scientist is sometimes so strongly biased by the hypothesis, that he tends to neglect "outliers (out-liars?)" that do not fit his hypothesised truth.

Moreover, what you are seeking to prove, you will often find proof for that. But you may not be aware that you have neglected other essential parameters or that methodological fallacies have crept in. If you would have tried to prove the opposite, you might have found proof for that too. If you are so lucky to realise that there are multiple possible ways to mathematically model and explain a set of data, so that you generate a number of parallel hypotheses, then it is still difficult to figure out which one reflects the underlying reality the best.

Scientists then often use Occam's Razor for this purpose, which states that the hypothesis with the fewest assumptions should be selected.
But that principle has neither been proven in any way, nor is it capable of revealing hidden assumptions in the hypothesis that seems to have the fewest number of assumptions. Perhaps if one would have known all the underlying assumptions, it turns out that this assumption was not the one with the fewest number of assumptions at all. In chapter 7 of part 2 I will show how this can indeed be the case.

Is science then a futile and precarious undertaking? Should we discard philosophy and science because they inherently cannot give us certainty?

What I have done in this chapter is shed doubt on science and philosophy to come to consistent knowledge about ourselves and the apparent world around us, which leaves no doubt. They do not seem fit for that purpose in an absolute sense, that we would be able know with complete certainty. But as long as they give us a pragmatic attitude and probabilities of likely success, a certain measure of predictability, via which we can make our lives more manageable and avoid misunderstandings, they are welcome to me. Science should be a pragmatic way of approaching reality with the purpose of creating technologies to improve our lives, but it should not be used as a way to come to absolute knowledge.

Just as science has pitfalls, so has Logic. If you know a little bit about the development of philosophy in the 20$^{th}$ century, you may have heard about the pitfalls of Logic that Bertrand Russell, Wittgenstein and Gödel have encountered. Via logic they have proven, that logic itself has a limited applicability. If logic is not the universal basis for knowledge, then what is? If we can't trust our overvalued rationality, what can we trust?

Logic bit its own tail with set theory known as **Russell's antinomy**[25]: Let *R* be the set of all sets that are not members of themselves. If *R* is not a member of itself, then its definition dictates that it must contain itself, and if it contains itself, then it contradicts its own definition as the set of all sets that are not members of themselves. Logic pushed to its extreme starts to contradict itself. In other words Logic has domains of absurdities, of paradoxes and even worse contradictions.
**Gödel** came with the **incompleteness theorems**. The first incompleteness theorem states that no consistent system of axioms whose theorems can be listed by an "effective procedure" (e.g., a computer program, but it could be any sort of algorithm) is capable of proving all truths about the relations of the natural numbers (arithmetic). For any such system, there will always be statements about the natural numbers that are true, but that are unprovable within the system. The second incompleteness theorem, an extension of the first, shows that such a system cannot demonstrate its own consistency.

Gödel's incompleteness theorems have a relation with the liar paradox (The semi-mythical seer Epimenides, a Cretan, reportedly stated that "The Cretans are always liars"). An example of this relation is the sentence "This sentence is false". An analysis of the liar sentence shows that it cannot be true (for then, as it asserts, it is false), nor can it be false (for then, it is true). Or what happens if Pinocchio says: "my nose will grow"?

A Gödel sentence *G* for a theory *T* makes a similar assertion to the liar sentence, but with truth replaced by provability: *G* says "*G* is not provable in the theory *T*." The analysis of the truth and provability of *G* is a formalised version of the analysis of the truth of the liar sentence.

So mathematical and logic knowledge, including language can never yield a complete and completely consistent framework.

And then finally Alan Turing showed that certain functions are not computable. This is also known as **Turing's incomputability**. It stems from, the Church–Turing thesis, which states that a function is algorithmically computable if and only if it is computable by a Turing machine. As there are functions that are not computable by Turing machines (see also the book Gödel, Esher Bach, an Eternal Golden Braid, by D.Hofstadter[26], herein incorporated by reference), there are functions which will always remain incomputable and to that extent also unknown.

In other words it has scientifically and logically been proven that knowledge, as we know it (i.e. knowledge which can be expressed in words and symbols), is not absolute and has its boundaries. There are boundaries to what can be logically and empirically known. Science and logic as we know them, are a nothing but a Cartesian house of cards built on quicksand.

Another positive point is that by philosophically realising that we have an experiential and interpretative bias, which gives us a subjective perspective on what happens, we can become more forgiving towards others. Others may have experienced the same event from a different angle, have different memories about the event (memories tend to fade and to transform over time) and most importantly a different cultural interpretation and different emotional experience of the event. As long as this is not clear to all parties, people tend to defend their "subjective truth", often based on fear based, territorial or social motives they are not even consciously aware of. Worse, in such a process people sometimes attribute certain "intentions" to the person they have a problem with. These presumed intentions are purely speculative. We don't know what is going on in the mind of somebody else. Even if accompanied by a certain body language, it is still interpretation. Unless you are telepathic it is guesswork you can better refrain from. When we realise that our "truth" is relative, we may become inclined to become more critical towards ourselves and more accepting towards others.

Yet another positive point of philosophical considerations as regards the working of the mind, is that we may start to realise that whenever we use (pseudo) rational arguments to defend a certain stance, these arguments are often driven by the wish to prove the desired outcome of the stance. That means that our selection of arguments is heavily biased from the onset. The most honest way to scrutinise a stance would be to start to find arguments and proof to defend to opposite stance. But even that is no warranty for success. As Wilson[36] already said, the "What the Thinker thinks, the Prover will prove". The "Prover" in us will find proof for what the "Thinker" thinks/desires. We are normally so strongly driven by our passions, that we have a blind-spot for the passion-driven selectivity as regards the arguments we provide. One may even question, whether we have free will at all; if there is ever any instance in which we overrule our passions. Because even if a rational argument would overcome the desire to appease e.g. a physical passion, one could argue that our passion for rationality at that moment has overruled us.

In a sense rational techniques when used for self-observation can be very useful, as long as we are aware of our potential blind spots. I have mentioned a few, but I suspect there are more of them, and obviously as they are unrevealed blind spots, for the moment I am not aware of them. Let's hope that the rational self-observation of my underlying motives will reveal further blind spots. Any suggestions as to further blind spots are welcome.

There is also the issue that if one wishes to enjoy and experience life directly in a state of mindlessness (or mindfulness as the Buddhists would say), one must have cleared out all the mental and emotional blockages that prevent such a state. As far as I know myself, these are usually the consequence of loops in the mind regarding unresolved psychological issues. You can only resolve such issues, if you are aware of them and if you are aware of your motives to allow them to persist. Whereas you can call self-analysis a form of psychology, the rational methodology you develop to do so is also a form of philosophy. It is not by trying to be mindless that you will reach a state of being mindless. The thought patterns that prevent the mindless state have to be worked out. In my humble opinion there is no better way than doing this exercise of self-analysis in writing. Writing clarifies the thought processes and makes your stance clear to yourself.

If you are a master of martial arts, music or art and you can work from that blissful state of mindlessness, which is certainly an advantage, both as regards the result and the enjoyment of the process of the act. But in order to become a master, one must go through a painful process of relentless practice. All the movements of sequences must have been automatised. It is usually only then that spontaneous improvisation will occur.

There are of course cases of prodigies who master skills without having learnt them. Also certain yoga techniques open areas where suddenly proficiency arises, without any trace in the practitioner's life of having learnt the particular skill. However, such occurrences are extremely rare. Even if such an emergent skill is attained by yoga, at least the practitioner has put in the required flying hours in the practice of yoga. That practice of yoga did involve self-study (svadyaya), which is again a form of philosophising. So practice, at least for the layman appears to be generally indispensable.

Now my ultimate target of this metaphilosophical analysis was to see if philosophising in whatever way is a way to come to correct knowledge, Pramana. Then we must see what Patanjali[23] means by the terminology Pramana. In Patanjali's Yoga Sutras I.7 we learn that Pramana is the knowledge obtained by either direct sensory observation, inference (deduction and induction) or testimony.

With regard to direct sensory observation, we must be vigilant that our observation is not tainted by sensory illusions (such as optical illusions) or other types of hallucinations. As soon as interpretation enters the game, there is a risk of arriving at "incorrect knowledge", which occurs when the mental concept and the sensory input do not match. If we can rule out sensory illusion, in such a case we can better question our mental concepts.

Deduction is one of the major tools of philosophy. *Prima facie* I would agree with Patanjali, that in as far as philosophising involves deduction, it appears to be a way to come to correct knowledge. I do not know whether the translation of the term "Anumana" (inference) correctly covers the intent of Patanjali to mean deduction by this word or whether Patanjali also meant inductive inference. As already said earlier, inductive inference gives a good likelihood of repeatability of a phenomenon, but no certainty. I strongly doubt whether Patanjali intended to include "inductive inference" this meaning. But even as regards deduction there is a problem: The premises of the deductive argument ultimately always derive from an inductive inference, which is never 100% certain.

As regards "testimony", one must be certain that the person testifying is a "truthful person". This of course is slippery ice these days. I most certainly do not trust the vast majority of religious texts, because they are full of internal contradictions. The only way here is by direct contact with a person or a text that you have not been able to nab on untruthfulness or internal inconsistencies. And even then there is the risk of wrong interpretation. It seems advisable to try out the teachings yourself to verify if they also apply to you.

The knowledge that you then obtain is according to Patanjali "correct knowledge". But we must still be aware that this is knowledge about how we experience the world. Our brain and senses filter information in quite an extreme manner, so that what is out there or the object of observation *an sich* (what Kant calls the noumenon) cannot lead to complete knowledge. We can extend our senses a bit with technical tools, but then we enter the realm of interpreting data, which is an unsure way to get "correct knowledge". Perhaps meditative techniques, such as "Samyama" (see Patanjali III.4), where subject and object merge can bring us almost complete knowledge of an object. I have a good hope that is indeed so, because as of yet Patanjali's Yoga Sutras have not revealed internal inconsistencies to me. But Patanjali never uses the word “complete knowledge”. In fact Gödel’s incompleteness theorem deductively shows that absolute “complete knowledge” is impossible. (Noteworthy, this contradicts the notion of omniscience of God as in western religions. However the Rg Veda, The Puranas and other Hindu scriptures do not claim omniscience of God. They state that even God does not know all his energies and is always enjoying discovering them).

One of the last questions I’d like to address in this brainstorm is: How can you ever know, that what you experience is not a form of hallucination? How can you be sure that your thoughts are your thoughts and not thoughts fed to you by a puppeteer? I pose this question not so much with regard to daily life experience, but more as regards the so-called mystical experience. I guess that as long as our experience does not enable, empower us to manipulate “apparent reality” the assumed mystical experience can have been a hallucination. If it does empower us we can still be the puppets of a puppeteer we’re unaware of, but in that case that distinction probably won’t matter to us at all. Like a little child watching a demo of a video game who has the impression he is steering the car in that game, it is probably very joyful.

As long as I am not a master in acquiring correct knowledge via meditative techniques, it seems philosophy is still part of my game. Utile instead of futile (within the “considerable set” defined by the limits nature sets to our senses and technological man-made sensors). But I am aware that my incomplete analysis may have been biased by the desire that this was the very outcome of the argumentation, that my argumentation may contain flaws and fallacies (please point them out to me) and that I have not sufficiently scrutinised the opposite stance and quantitatively weighed the different opposing arguments in a balance.

Here I stop my brainstorming, and promise to work out a personal philosophical methodology in more detail, that allows for a fairer scrutiny of the opposite (the futile) stance. One thing is sure however: we can never be sure that we have all knowledge to come to a fair balancing, so that it seems as per Gödel’s theorem and as per the blind spot to be able to see all possible vantage points, that the issue is ultimately undecidable.

This notion then prompts me to continue to pragmatically apply my philosophy as long as I have no good reason not to do so. In chapter 2 of part 2 of this book I will show you how philosophy can be rendered even more utile by confirmation with Technological applications, that can show that within your "considerable set", your premises have a reasonable predictive validity. We may never know what is the ultimate "Truth", but at least our science and philosophies are utile in as far as they can be technologically applied.

## Chapter 6 A new Measure of Order and Chaos

In this chapter I will question the present day understanding of the notions of order vs. chaos and entropy vs. syntropy (or extropy). I will propose a way to quantify syntropy and try to show that order and chaos are mutually dependent and hence different sides of one and the same phenomenon.

### Background

Scientists consider the possibility of a so-called heat-death of the universe: The second law of dynamics dictates that the entropy (degree of disorder) of a system must always increase for a process to occur. At a certain point when everything has drifted apart such that the temperature differences of processes can no longer be exploited to perform work, entropy can no longer increase and nothing can happen anymore. This is of course a very miserable prospect for the future and fortunately there are different scenarios envisaged, because the universe is not necessarily in an "open" or "flat" configuration. The notion that the universe will always expand appeared to have been proven in 1998 on the basis of red-shifts in the spectra from distant galaxies. Based on this notion it was supposed that the universe will end in a big rip, big chill or heat death. Today, however, the notion of an ever expanding universe is questioned again, and the ideas of Frank tippler, that at a certain point the universe could start to contract again, could perhaps be reconsidered. But that's a topic for a different essay.

This essay will try to convince you that the present day understanding of entropy, extropy, order and chaos may be incomplete. In my philosophy these opposites balance each other. Nature's intelligence algorithm strives for integration of functional aspects and in doing so it creates diversity and complexity. This integrative functionality in my philosophy derives from the fact that every self-sustaining system has an inherent kind of integrative information feedback mechanism, which we could call a form of consciousness. In other words self-sustaining system natural systems (such as living systems) are animated. This is a particular type of panpsychic philosophy, which does not attribute consciousness to artificially compounded systems such as a bimetal in a thermostat or a rock, but only to systems which naturally evolved and were capable of sustaining themselves and if needed adapting themselves to the environment via Nature's intelligence algorithm.

In this essay I will propose some new concepts which hopefully may one day lead to a formulation of a measure for syntropy (a.k.a. extropy or emergence).

**Order versus Chaos**

In Nature there is no perfect order, since if it was, the atoms would lie completely still, not vibrating at all at the hypothetical "absolute zero" of 0°K or -273,15°C. But this temperature is purely theoretical. There is no existing system which is completely at rest. In other words every natural system has a certain degree of order and a certain degree of chaos. Inherently and per definition.

Verlinde's entropic gravitation is based on the notion that if similar information content clusters in such an informational matrix this has an advantage for the dissipation of undifferentiated energy. In other words by creating local order, overall entropy can be increased. Chaos is optimised by local levels of order. Chaos cannot exist without levels of order nor can levels of order occur without the creation of an ever increasing chaos. In that sense Chaos and order are two sides of the same coin, giving rise to a fractal of ever newly occurring levels of metasystems. In natural systems every level of Chaos has an internal order, which repeats itself over scales giving rise to a chaos-order fractal; every order gives rise to chaos. A great book on this topic is "Chaos" by James Gleick. The ordering principle integrates the most basic elements of each level, which when compounded generate new basic entities of a higher meta-level. These entities can be recombined and integrated in all possible permutations in a computational screening and pruning process, which creates a new spawning of all kinds of combinatorial forms. Thus integration and differentiation are also intertwined concepts.

There is also a phenomenon similar to the Braess paradox, which shows that if you continue to add more and more order to a system, at a certain point chaos emerges and no further gains in decreasing entropy can be obtained.

In other words Chaos and order cannot exist independently of each other, which means that they are balanced in a certain way. Our mental and mathematical representations of perfectly ordered systems have no physical counterpart in Nature.

**A measure for Syntropy**

In this chapter I'd like to formulate some brainstormed ideas about quantification of order as imposed by self-generating, self-organising and self-sustaining systems i.e. autopoietic systems. This chapter is a bit mathematical and you can skip it as five tons of flax in Discordian terminology if this is not your thing.

It is an exploration into how a panpsychic (and even pancomputational as we shall see in part 2 of this book) substrate is capable of self-organising against the entropic forces. The ideas are purely speculative and could be labelled as pseudo-science. Personally, I see it as an attempt to set aside a prejudice against Panpsychism and to reason with formulae, if we can come to a better understanding of order and emergence.

In this attempt I would like to integrate the information on emergence, self-organisation, teleology and autopoiesis (self-generation and self-sustention, literally "self-making").

Entropy does not seem a good measure of order/chaos to me, when it comes to autopoietic, self-organising systems (e.g. living systems), which can replicate. Living systems have different semi-autonomous levels, which are separated by so-called meta-system transitions. Ordering seems to occur by resonating to a pattern.

Each level has its own **ordering capacity** contribution, its own **resonance <u>recruiting</u> frequency**, which must be multiplied by the number of entities in the level to get to the total ordering capacity.

The order of a system is not simply its information-wise statistical orderly spread in an n-dimensional grid, but also should entail the capacity to autocatalytically regenerate itself, which I call its "meme-spreading capacity", its "virality/infectivity" or its "proliferability". Memes are ideas, paradigms or behaviours, which spread from person to person in a culture.

As far as I know there are not many theories describing the virality/diffusion speed of meme spreading.

Autopoietic systems exist in a hierarchy of levels, where the number of autopoietic entities encompassed by the level above increases when one is going to higher levels. A set of examples, from the lowest physical systems up to the higher "living" systems that we know of, is presented hereunder. On the right hand side is the number of units of the sub-level entities in the higher level entity:

| | |
|---|---|
| Get 1 photon out of pure chaos | unity |
| 2 photons -> 1 electron, 1 positron | $1\text{-}10^1$ |
| 3 quarks -> 1 proton or neutron | $1\text{-}10^1$ |
| 1-100 electrons +1-50 protons/neutrons -> 1 atom | $1\text{-}10^2$ |
| Molecules: amino acids, nucleotides, sugars, lipids: | $10^1\text{-}10^2$ atoms |
| | |
| Macromolecules: 10-10000 monomer units | $10^1\text{-}10^4$ |
| Bacteria | $10^4\text{-}10^6$(lipids+proteins) |
| Eurkaryotic Cells: | $10^7\text{-}10^9$ macromolecules |
| Organism: | $10^{10}\text{-}10^{12}$ cells |

(Cell organelles or organs are not independent).

There is a more or less logarithmic increase in the maximum number of sub-constituents going from lower to higher levels.

Every new level of autopoiesis (self-sustention) is a new level where the subconstituents seem to have "agreed to cooperate" and seem to have defied the forces of an inherent selfish nature. But they can do so because of the emergence of a common morphogenetic field (in the sense of Sheldrake's[27] morphogenetic resonances) which may direct them: A higher level/sense of resonance.

These collective resonance fields may not suffer from the ultimately reductionist approach of the weak emergence (weak emergence, in which the emergent property is reducible to its individual constituents; strong emergence, in which the emergent property is irreducible to its individual constituents).

A newly occurring collective resonance could be considered to constitute a new entity of “Spirit” in a philosophy of Panpsychism; an Anima, a Soul, a Purusha.
Although such a spirit is ultimately in- or transfinite and a whole, for the sake of reasoning, each fractal level is here considered as a quantum, or collection of quanta of spirit. I do realise that ultimately everything forms a whole and that the consideration of quanta of spirit does introduce an undue element of reductionism, but we cannot reason without it, because our language is not dynamic process oriented as in Bohm's Rheomode[14] (cf. “The Implicate Order”), but rather “object oriented”.

To compose the formula for the total autopoietic ordering potential Psi $\Psi_{tot}$ we may need the following elementary representattions (a hash # symbol stands for “number of”) :

#level1 entities = #LE1 and generalised: #level n entities = #LEn

Virality or Resonance Based Infectivity (RBI): This is the spreading capacity of level-entities via resonance: $RBI_n$, in which n represents the number of the level.

(This RBI also tells us something about the vibratory frequency of the entities, a kind of panpsychic Eigenfrequency. Each level also has its weak and strong **Emergent** attributes: But these are subsumed in the RBI: see below).

Extropy $1/S$ is here defined as thermodynamic/statistical spatiotemporal spreading ability towards order (the lower the entropy (provided that it is above an immobile i.e. solid level), the higher the chance of cooperative aggregation).

The ordering potential of a level n, $\Psi_{n,}$ I propose to be the product of the number of entities in that level ( #LEn), their Resonance Based Infectivity at that level ($RBI_n$) and the extropy of that level ($1/S_n$).

Considering unlimited resources this formula must look something like this per level i:
$\Psi_i$ =( #LEi x $RBI_i$ ) / $S_i$

The total Ψ for all n levels together is then:
$\Psi tot = \sum_{i=1}^{i=n} \Psi i = \sum_{i=1}^{i=n}$ (#LEi x $RBI_i$ ) / $S_i$

#LEn and Sn are quantifiable. The difficulty lies with RBI.

Each system has its own idiosyncratic **emergent** features (features present in the system as a whole, but not present in each of the building blocks when taken alone), which promote their RBI. For instance consider the simple example of a "water" molecule $H_2O$ compared to its building blocks O and H. When compared to its sub-constituents, it shows emergence of geometrical aspects (tetrahedron), which are a consequence of the new electronic distribution. This has further effects on the reactivity (and infectivity) of the new holon, in the sense that it can make Hydrogen bonds with other water molecules and shape the space around it, and even exchange H atoms between molecules (acid-base reactions). As of yet I have not found a way to quantify the "Emergence". Goertzel also attempted to formalise "Emergence" in "Creating Internet Intelligence" [13] and "The hidden pattern"[28], but as of yet he has not been able to come with a quantifiable solution either. It is the lack of a good quantitative theory of "emergence", that makes that "entropy" and "extropy" as we know them are imperfect measures of chaos and order, because they do not capture the concept of the teleological ordering attractor, which brings about order by spreading new swarms of resonant infectivity. Sound and waves may not only serve as a metaphor, but may be the ultimate forming agents, functors shaping structures. Shakti, Ki, the dynamic principle longing to reunite with her ever all-penetrating heavenly consort Shiva, the seemingly static all-pervading consciousness.

Higher levels have more/other versatility: e.g. self-folding of proteins according to well-established chreodes (necessary pathways). Receptor ligand interactions allowing for conformational changes etc. etc.

But essentially the contributions in the example are of a morphological and electrondensity based reactivity at what is still a molecular level.

Above this level (cells, organisms) we get higher order signalling, increase in number of adaptive states of a system, sexual reproduction over asexual reproduction, social/group limitations in the form of behaviour, (between entities that spread ideas that enhance their adaptability, one can even speak of a meme-infectivity) etc. and it becomes more and more difficult to describe the Resonance Based Infectivity of a system. The higher the degree of self-organisation, the more emergent properties arise, the greater the meta-vari(eg)ation and adaptability.

An ultimate theory of orderliness would also need to take into account the environment and limits imposed thereby on the Infectivity/Spreading ability of a system.

**A Measure for Natural Intelligence**

My prior hypothesis was: The degree of intelligence of a (sentient) system can be expressed as the product of (meta-)vari(eg)ation and order.
I'd like to shift the accent of this idea. The degree of intelligence of a (sentient) system is probably **proportional to** its autopoietic ordering potential Psi.

The ordering potential also comes to expression in the form of adaptability, which entails its capability to generate meta-vari(eg)ation but it also entails Perceptibility. Perceptibility must be a measure of the variety and intensity of signals, which an autopoietic system can process: Perceptibility is therefore a product of the perceptional variation potential and Weber-Fechner's law of Perception. Adaptability arises from the capability to change between strategies and from the capability to generate strategies (creativity). There must be a kind of optimal adaptability, since a system that tries too many changes or screens too elaborately is not effectively adaptive but loses itself in the game of providing alternatives. A highly intelligent system has effective pruning strategies to rapidly discard potentially ineffective strategies.

The more alternative pathways are available in a network (such as e.g. a neural network), the more possibilities there are of resonating with some signal from the environment.
A hyperconnected network in which all the subconstituents, the nodes, are linked to all other subconstituents only yields meaningful and reasonable strategies, if the links are not all the same. However, by attributing different weights to the links between the nodes, different patterns of resonance can occur. The more patterns can be perceived (i.e. the higher its perceptional variation potential is) and effectively pruned, the more efficient the system. Adaptability is possibly strongly connected with the ability to effectively prune, which ideally is an optimised process. Too zealous pruning may discard potentially helpful strategies; of course this is a looped process, where if the most promising strategies fail, screening of less promising strategies will be attempted.

The pruning resulting in a been-there-done-that attitude towards non-effective strategies optimally will favour cooperative harmonising systems. Firstly because that's the Nash equilibrium and in the long term the most advantageous solution to e.g. Axelrod's "Prisoner's Dilemma" (two prisoners can choose to cooperate and both serve a short period of incarceration and not betray each other or cut a deal with the prosecuter, resulting in a shorter incarceration for the betrayer but a longer incarceration for the betrayed prisoner). Secondly because the harmonic resonance resulting from that strategy is the best spreading meme resonance. Axelrod's optimal "Tit for Tat" strategy automatically results in a natural "friendliness" and a forgiving nature of the system. The cooperation is ultimately in servitude to the higher level entity.

The RBI (Resonance Based Infectivity) of the autopoietic ordering potential therefore entails Perceptibility, Adaptibility and the Emergent means of communication with the environment (corresponding also to the input-, throughput- and output-processing abilities; cognise, experience, recreate). It is possible that this can all be described in the number of nodes, the nodality (connectedness) and the weighted signals between the nodes. The nodes are the entities constituting the levels. So you end up with a kind of optimised multilevel neural network-type of system as the ultimate processor of resonances.

The autopoietic system is not to be reduced solely to a brain as we know it though; I presume that all levels, from electron to organism, all transmit information from below, which is integrated at the brain level. And conversely the autopoietic system receives resonances, signals from its environment; from above if you wish. Between the entities at the same level there is more interconnectedness than between entities at different levels.

But there is also cross-level communication in the form of special signalling molecules such as hormones, transcription factors etc. Key to the Perceptibility, Adaptability and the Emergent means of communication, is the concentration of energised awareness to enliven these functions, which cannot be solely expressed in terms of the nodal network and its flow patterns; rather it determines the speed with which these patterns can be purposefully accessed.

The processing speed is a strong indicator of this aspect. This aspect of energised awareness resembles the hypostatic (i.e. underlying primordial) consciousness of existence, but then in a concentrated and directed form thereof. In fact it may well be that the thusfar described functional and structural elements are the mere manifestational reflection of the amount of "hypostatic spirit" that has been concentrated/personified.

So the manifestational aspects do inform us about the concentration of awareness in the system and give us a kind of relative measure. But it is impossible to know its absolute calibration point; there is no absolute quantification of the autopoietic ordering potential Psi possible in this hypothetical framework.

**Beyond selfishness**

I postulate that when the Psi increases beyond selfishness, the system may be teleologically attracted towards a more universal degree of harmony with every other entity, resulting in what the Greeks called Agape or divine Love.

To establish this, the system will need to send out a frequency which fits and resonates with all the other frequencies. This leads to the idea that this must be a very high frequency so that whole number multiples of this frequency fit into any other frequency.

It is here upon transcending selfishness that paradoxically for the first time real individuality can be experienced (or should I even say "enjoyed?). When selfishness is transcended the system works for the greater good of the whole metasystem of which it forms a part. As the system now operates in resonance with the other parts, it is no longer in competition and can save the resources it used to spend on its competitive resource allocation. This gives it a new freedom to specialise in a certain function, namely the "Dharma" of the system where its natural proficiency lies. In fact a part of its normal level "freedom" is sacrificed or outsourced to the higher meta-level, allowing it to evolve its qualities and build new relations defining its new structure.

It is at this point that the notion of "autopoietic" is transcended. Because autopoietic in a certain sense implies being a-teleological: having no other purpose than maintaining or continuing their existence. I once read the following: "Imagine a bubble. An autopoietic system is a bubble in which there is only ever interaction, reference, or communication with *other elements **in*** the bubble, never directly with other things *outside* the bubble".

Upon the meta-system transition described above, where parts choose to operate to form a new emergent whole, which is more than the sum of parts, a higher teleological purpose can be served. New interactions can be engaged with new entities on a different meta-level. A "brane" (or membrane) so to say has been formed among the constituent system parts, which now form the new meta-system. And automatically a new "currency" of interaction applies: the currency of e.g. hormonal signalling between cells, now has become language and money between people. A currency that naturally emerges and which apparently cannot be comprehended in terms of the underlying physical elements, but which in fact metaphysically resonates over the (mem)branes; the onion shells of the preceding noetic event cascade. A cross-telic feedback over meta-levels and over time, that reverberates the "seemingly fossilised" feeling structures (e.g. tissues; a terminology suggested by Eric Ryser a.k.a. Eyeorderchaos[29]) and currencies of lower levels. Where mappings build new structures and structures generate new mappings, which as a Yoneda-embedding fractal are ever reminiscent of the underlying oneness from which they originated.

The new meta-system has several kinds of collective advantages, which basically obey the principle of the Nash equilibrium: What is good for the whole is good for the individual, yes even better than it could ever have been on an "individual level". It is here that when "Egoism" is transcended for the good of the "hivemind" (as synonym for meta-system), true individuality can be empowered in a "meaningful" manner. Remember that meaning ontologically occurs as a di-density: it takes at least "two concepts" or in panpsychic terms "two entities" to establish a "relation", which confers meaning. Alone is stupid, together is smart.

Unfortunately, mankind with its obsolete separatist competitive and dividing economic and political systems appears to be unable to let their "reason" transcend the pettiness of "autopoietic egoism". We still haven't grasped the ultimate truth of "optimal integration". Cooperation, if performed on the basis of mutual consent largely outperforms competition. I am not advocating communism here, because that is a tyranny imposed from above without the mutual consent of the parts. But a social democracy of "true individuals", "self-realised people" or "adepts", who resonate with the whole, without being dictated by it. Present day capitalism with its ever demanding imperative of "economic growth" only warrants a future Malthusian crisis (Mass starvation as a consequence of exhausting food supplies). An exhaustion of the resources as a consequence of cancerous growth of parts, which were unable to cooperate for the greater good of the whole.

As described in the recent Tegmark[30] paper on "Consciousness as a state of matter" there is an optimal balance between integration and autonomy of the constituent parts, which also optimises the information stream density between the constituent parts. If you deny too much autonomy, the system will become static, but if you give too much autonomy the system also grinds to a halt (known as the "QM Zeno-paradox").

These (meta)-physical laws of integration are "Pantelic", they apply to any teleologically attracted system. Any higher intelligence will be able to apply this ever renewing imperative of meta-system transition as the ultimate guarantee of survival of functional memory of mappings in ever changing paradigms of structure.

Noteworthy, these processes hint towards a pancomputational nature of every level. In part 2, chapter 1 we will see how pancomputationalism may lie at the heart of every form of existence and even at the heart of consciousness.

**Conclusion**

You have seen how the concepts Syntropy and Entropy as measures of mutually dependent order and chaos were discussed in this chapter in the framework of emergence, information and higher order systems, which may require a different quantification if we include consciousness in the equation.

## Chapter 7 The boundaries of Infinity

In this chapter I will challenge the present day interpretation of Everett's many worlds interpretation (MWI).

### Background

In 1957 Hugh Everett came with a very original alternative to the so-called Copenhagen interpretation in quantum mechanics. In the Copenhagen interpretation before a measurement is made a quantum system is present as a superposition of probabilistic wave functions, but only upon measurement the set of probabilities is reduced to only one of the possible values. This feature is known as wave function collapse.
In Everett's MWI a wave function collapse never takes place, rather at every event a universe branches into multiple parallel universes, each of which shows one of the possible values of probabilistic states upon measurement. In the so-called "Schrödinger's cat" example, in which a cat in a box is in a superposition state of dead and alive, opening the box as measurement will collapse the wave function into a living or a dead cat in the Copenhagen interpretation, but in Everett's MWI the universe will branch into two universes: One in which the cat is alive and one in which it is dead.
Today many physicists who prefer Everett's MWI over the Copenhagen interpretation, have gone so far as to consider that Everett's MWI means that everything that possibly could happen actually happens somewhere in a given parallel universe. An infinity of worlds is generated.
One of my friends called this interpretation "Infinitism".This is a bit an unfortunate name, because according to Wikipedia "Infinitism" is the view that knowledge may be justified by an infinite chain of reasons. It belongs to epistemology, the branch of philosophy that considers the possibility, nature, and means of knowledge.

How unfortunate that the philosophers have claimed this very finite definition of a terminology, which seemed to fit so well Everett's MWI, which could have implied so much more...yes endlessly more.

**Infinitysm**

I'll try propose a new all-encompassing indicator for this notion and I will use the terminology "Infinitysm" instead, which cannot be a definition, because the Infinite cannot be finite. An attempt, which is bound to fail, because the Infinite cannot be grasped by a finite set of words. But a good description may give a hunch of what I feel it points to.

For me the notion of Infinitysm would relate to the concept that any aspect of existence would be infinite.

The Infinite, however, cannot be defined by anything. And as it has no limits it must necessarily encompass all what is, for if it were to exclude anything, this would already impose a limitation on the concept.

One could argue that it also implies that anything which can be possibly thought of, must de facto exist within the view of infinitysm. Any scenario of what possibly could happen to you does then happen somewhere in a parallel universe or somewhere in the infinity of time. Like Boltzmann Brains popping up in the Sea of Dirac. Tegmark pushed to the extreme. This is Everett's[31] multiverse of an infinite amount of parallel universes. What can be thought of, must necessarily be there, according to these scientists. Including a plethora of hellish nonsensical universes, where we repeat the same actions ad infinitum or commit cruelties all the time.

If one starts from another perspective, it could be said that what is "infinite", is that what is not finite. Since every object and concept is physically and ontologically, respectively, delimited, it seems a strange hypothesis to assume that there can be infinity in collections of finite things. At best there could be a transfinite number in sequence of them over time, if they continue to be created. Transfinite numbers mean that there is no end to this sequence of numbers, but it does not mean that all aspects of the phenomenon or object are infinite.

An infinite collection of all possible things and objects must mean that all aspects thereof are infinite. It implies that everything is already there and time and creation ultimately do not exist. This appears contrary to our observations.

The infinite-collection-type of infinitysm runs into logic contradictions. Because a thesis and its mutual exclusive opposite must both be able to coexist... and yet not, because the concept that they do not coexist, must also exist within infinity. If infinity has an aspect of suchness, it would be delimited, but if infinity excludes an aspect of suchness it would also be delimited, by virtue of the absence of the aspect.

Logic is about a world of here and there, now and then, about polarities of any kind. Logic deals with a world of things. In other words logic is a set of rules for a system built on dualities. Logic compares specific instances with general rules. Which does not mean that those rules cannot have exceptions.

But the very nature of existence might be non-dual. Everything being infinitely tangled up. A supermetatautological holographic fractal-network of self-similarities, where only limited perspectives of a confused mind believe that there are "things" with a given “space-time location”.

The very nature of being may transcend logic as we know it. Because maybe there is only a single "whole" of reality and any delimitation of a concept, any perceptual delimitation and classification of a perceived object as being "part" or belonging to a category, are by-products of the workings of the Mind and the Ego. The confusion of us as the seeming parts is a result of our belief, that we are separate from the whole.

Infinity must therefore transcend logic and mindlike comprehension. Aspects of suchness and their absence are both part of it. How? Perhaps by the concept of time and space? Infinity encompasses all times and all spaces thereby encompassing all aspects. But not every seemingly partial spacetime universe may encompass all aspects in a manifested way (though it does in a potential way).

But what then about the concept that every seemingly partial spacetime universe needs to encompass all aspects in a manifested way in order to fulfil the requirements of Infinitysm?

**Religions and Transcendence**

If this would not be possible, then infinity would be limited by an impossibility.
Again it is logic biting its own tail, like the Ouroboros from Gnosticism. Infinity cannot be grasped in words and logic. Because there are no parts. Parts are illusory. So then infinity would be delimited by the notion that parts are illusory.

The only explanation can come from transcendence. Any notion is there and yet not, depending on the perspective. As Nietzsche said: no way of seeing the world can be taken as definitively "true". (Hence my allergy to belief systems, which claim to know "absolute truths" about how we should interpret the relative world around us. In this relative world of physical and informational existence, every "thing" is relative. Although the meaning of this statement itself implies an absolute truth, this is not even self-contradictory, because this cognitive meaning we recognise with our consciousness is not a "thing", -not even information- but an inherent aspect relating to the whole of reality. Absolute truths can only be aspects of the whole of reality and its underlying ground of primordial consciousness).

True and false fading into each other like pictorial values of a kaleidoscopic spectrum of Perspectivism.

So perception is merely a learned way of connecting the dots in a certain way, where other ways of connecting the dots are also possible. The cliques formed in the grid of your mind are not necessarily the cliques existing out there...and yet not.

Religions claim God to be infinite. From a logic point of view it would appear that this cannot be an absolute infinitysm, but rather a good attempt to approach this. God can then be transfinite in many aspects. But any system that tries to represent all perspectives eventually runs into limitations of Pinocchio paradoxes: What happens if Pinocchio says: "my nose will grow"? Gödel's incompleteness, Heisenberg's uncertainty and Turing's incomputability show that materialistic conceptualisations have their inherent limits. The claim from religions about infinity originated at a time that the notion of "transfinite" had not been defined yet. Had they known about "transfinity" perhaps they would have claimed God to be transfinite instead of infinite.

To transcend these, a different type of non-exclusive holistic logic is needed in which there are ultimately no parts, but only wholes at different fractal levels. Where the matter of the universe (multimetaverse) is but a transient spiritual expression in the mind of (a) God; a snippet of consciousness. A continuously changing self-resolving Paradox. Every seemingly self-similar soul starts on the rim of the fractal of consciousness as a seeming part unknowing of its unity and interconnectedness with the whole. As it grows, it becomes more and more self-similar to the Ur-form of (a) God's morphic resonance, and by resonating it becomes conscious and aware of its unity with the whole. And yet not, as there are also other ways of navigating the fractal of infinity. So is it technically possible that "ultimately we'll all return to (a) God"? In terms of transcendence, the answer must be yes and no and "Mu"(undecided or both). If after every time your body dies, you return to the "Omega endpoint" (Teilhard de Chardin[9]), you can explore other dimensions from there. After a series of reincarnations in different dimensions of a spiritual nature, after a series of levels in a computer game, after having spent some time in this heaven, you may have to return to the material Matrix. So "ultimately you return to the matrix" may also be true and yet not; is there an infinite loop between matter and spirit?

Is the nature of existence limited by any concept? From whatever dimension or level? Are there "Pantelic" laws, i.e. laws that are valid in any aspect of manifestation? If there is a God or all-encompassing consciousness, does it have an ultimate Ur-form, Svarupa or Suchness from which all forms and things are derived? Are all conceptual theses and antitheses ultimately resolved in the Omega point? Are the Platonic world of archetypes and the Aristotelian vision of animism at all levels of existence simultaneously true? Are all Theses ultimately resolved to be true and false simultaneously by existing in parallel? So that all philosophies and religions are ultimately true somewhere, somehow in a parallel universe? So that the universe is ultimately both the retro-causal teleological simulation product of a <u>material</u> hypercomputer in the Omega point (McKenna's "Eschaton") and of a spiritual non-material God, in which matter is but a thought form? If a God exists, is this God omniscient and omnipotent, or do these concepts carry an inherent paradox? (God cannot make a stone so heavy that he cannot lift it; If God is and knows all dimensions, all times and all multimetaverses then the whole of God would appear to be a non-evolving collection of frozen eternities and not an entity which experiences). Is there an ultimate transcendent state that encompasses all points of view (like the Elephant in the Buddhist tale which is shown to a number of blind men, who all touch a different part of the elephant and then describe it all differently and start to argue what an elephant looks like, whereas for a seer there is one transcendent entity, the elephant)?

If not, what then is this strange form of limited consciousness that I call "I". If not, will we meet again? ... or yet not?

Jesus said: "You cannot pray to both God and Mammon". One of materialism and spiritualism must be ultimately wrong as ground principle of existence. One must encompass the other which is merely an illusion. For me ultimately there must be some kind of logic, even if it is not logic as we know it.

**Spiritualism**

I place my bets on spiritualism. Because of emergentism. Because the whole is always more than the sum of parts in unexpected ways. Because the cosmos at every level evolves. And because I feel this to be true, because it resonates in my heart. And because of this, so-called presupposed infinity of manifested existence at each and every given instant must ultimately be limited. Has ultimately "a suchness" (Tattva). A suchness, an aspect of evolving intelligence. An aspect of harmonious solutions giving rise to stable resonating forms. Because by resonance those manifestations, which were temporarily lost by apparent destruction, can re-occur. Because being, energy, consciousness cannot be destroyed. So that there is ultimately a pragmatic "Goodness" and sustainability/recurrence of "sense" (in the sense of the ability to feel, to sense: sentience).

Because there are too many coincidences of patterns in our solar system, the measures and locations of the pyramids and in the size of the Earth and the Moon (the Man-tra of Moon and Terra), which point to a beautifully fine-tuned design by a higher intelligence. Information must have come from above or beyond to those who invented the "mile" and the "metre" to precisely encode the "Heavenly Jerusalem" (see J.Mitchell's[32] "Sacred Geometry") and the speed of light. In fact the mile and the metre are a kind of Rosetta's stone, a key or "Clavicula" to decipher the presence of higher intelligence. The question then remains, how did the information of the mile and metre arrive in the minds of the people who are credited to have invented them? Were they influenced by this higher intelligence? Are entities from a Kardashev IV society influencing our minds?

I place my bets on spiritualism because of synchronicities and cross-temporal information transfer -as in the television series "Touch"- that we observe.

And because I am essentially here in a given conformation and form, a suchness, which must therefore influence and codetermine the rest of the manifested whole to which it is connected.

Because there is ultimately a continuum: Because emptiness and separation would imply non-existence, which cannot be. A continuum the medium of which is primordial consciousness,

In the phrase above, by what I called "existence", I meant the sum of both matter and non-material physical presence such as energy, fields, the ether of Akasha etc.

On the other hand one could also define existence as that what "stands out" (and can be directly perceived), which would limit this to material manifested existence only; Ex-Sistence. The energetic non-manifested energies which serve as homogeneous background for the existence are then Sub-sistent.

It is my guess that as a seemingly individual consciousness grows, it learns to tune into ever finer frequencies and it will perceive ever more so that its definition of what is existence thereby changes.

But the question remains: Ultimately, on the level of what we would consider to be the "God level", is the whole of existence perceived there in all its details? (Even God does not know all its energies... says an ancient Indian Vedic scripture...defying omniscience).

I guess even the highest God needs a kind of time and a lack of omniscience and omnipotence in order to be able to exist or subsist, in order to be willing to exist or subsist?

So transfinity probably exists if there is no end to time. A somehow limited God with a huge knowledge and power as regards future potentialities like an entity from a Kardashev IV society is not unthinkable. But an absolutely infinite entity, in the sense of Everett's multiverse[31]? I doubt it.

Existence is defined. It has the boundaries of not being nothing, of not being infinite. Infinite means no boundaries, no definitions and hence no suchness, no qualities.
Existence has qualities. It evolves. It is essentially benign. There is more constructive resonance within existence than non-resonant interference: Otherwise we wouldn't be here to observe existence. It appears to have constants and laws. Even if there would be a multiverse encompassing all kinds of possible universes, these universes must at least be possible. They need certain laws and constants. If all constants would constantly change, they would not be constant and no structures could arise.
So if there is something/someone like a God, it probably does have attributes and qualities, namely all those we know from our observations of existence plus some more which we may not yet know. Qualities that have been fine tuned to provide Beauty.

The Art of Beauty is to leave out the superfluous and to keep the essential. In music a great musician can tell a beautiful story by omitting notes at the right moment. Beauty is often not more "quantity" but rather more “quality”. There is an inherent beauty of parsimony and a parsimony of beauty. As we observe that our universe is profoundly ordered and beautiful, how can we believe in Everett’s[31] plethora of cancerous universe-multiplication ad infinitum? Is there no screening and pruning of those universes that end up in endless loops? Universes which can no longer evolve? Universes that have turned into a Hell, by virtue of their repetitiveness? The Intelligence algorithm of Nature seems to eliminate its non-functional strategies.

Sens, Sense, essence, and presence all have certain attributes of constantly changing (“*panta rhei oudi menei*”) input leading to output, which makes that essentially any living soul is a kind of computing device, although we may not know its algorithms. In that way Pancomputationalism can be united with Panpsychism.

Each time an energy, a particle or a soul resonates, communes or rings, the new pattern still fits within the resonance envelope, but also changes it slightly (but never in a divergent way). Pan-resonantism, Brahma Nada (the sound of Brahma. Funny that in Spanish "nada" means "nothing"...). The most essential quality of existence is "Love" as it is a communion or resonance of energies, particles or souls.

In terms of interpersonal attraction, four forms of love have traditionally been distinguished, based on ancient Greek precedent: The love of kinship or familiarity (in Greek, *storge*), the love of friendship (*philia*), the love of sexual and/or romantic desire (*eros*), and self-emptying or divine love (*agape*).

The love of kinship or familiarity or friendship has mostly to do with resonance of sameness. Romantic love involves desire and must sprout from a lack of completeness and hence from a difference. Both involve attraction and attachment. Agape gives up the Ego and is considered as the return of Love to God. Attraction to the Godly.

It seems that it is by attraction that structures can form. No thing appears to exist without attraction.

The Absolute, the infinite appears basically devoid from qualities. (A) God as all-encompassing sensing entity is not. If we consider “God” to be “Consciousness”, it is Good, because this God wants to perpetuate and cannot do but that. This consciousness-God IS then quality, noumenal Sense, that brings forward apparent manifestations, phenomena. Non-being is not possible in this perspective.

Nature shows the way: in Beauty and Bliss. Destructions are rare but do occur (e.g. the dinosaurs destroying meteorite) to allow higher conscious forms to evolve, when evolution stagnates.

The transcendence of the duality determinism-indeterminism can be found in the analogy of a fractal, whereas the specifics on the rim are indeterminate and always yield new structures; at every level the same laws impose the same general structure or determinism.

In the penultimate chapter of part 2, we will see how these preparatory ideas find their apotheosis in a reasonable framework of what the “Highest Transcendence” could possibly encompass.

**Conclusion**

I hope I have been able to illustrate to you, how everything belonging to the material world is likely to be limited and that it is unlikely that there is an infinite collection of multiple parallel universes at any given moment. We must carefully investigate to what extent Everett’s “Multiple World Theory” can be reconciled with notions about infinity from science and religions.

## Chapter 8: Find your Soulmate

Memetic alignment for psychological profiling

### Background

Imagine you're on a social medium or even a dating site and you're trying to find people who share the same interests as you do. Usually such sites have some fields where you can specify your interest, indicate what books or films you like etc. If the site has a good algorithm it might indeed search for an optimal alignment between your interests profile and that of others to identify potential friends or partners. However, not everybody takes the time to fill out such questionnaires and due to this lack of accurate input the results may be less good than you expect.

There are however other sites, where people make electronic collections of their interests, such as *Tumblr*, *Stumbleupon* and *Pinterest*. *Pinterest* is a very interesting site for determining someone's psychological profile. On *Pinterest* people make collections of images (including images with accompanying texts) of topics they are interested in or like. This is excellent for psychological profiling, because it gives a great visual overview of all interests in different categories a person might have.

If *Pinterest* could be improved and enhanced in its functionality so that you could compare the degree of overlap in pins you have with somebody else, in principle you could be able to find people who share a maximum of interests with you and who could be potential friends.

### Memomes

In fact, your *Pinterest* collection is in a certain way your *Memecard.* Memes are elements of sets of beliefs, behaviours and ideas, which are typical for a certain cultural group and which can be easily spread among people. Religions for example are very strong memegroups or shared *Memomes*. But also a political or sportive affiliation can make that you belong to a certain *memegroup*.

Each idea category typical for such a *memegroup* can be called a meme, in analogy to a gene.

A *Memome* can then be considered to be a subject's complete set of memes in analogy to a complete genome.
In genetics if you want to determine the degree of similarity (homology or identity) between two organisms, you carry out a gene sequence alignment.

In analogy, an algorithm could carry out a "meme alignment" protocol on *Pinterest*. Each image shared by two different individuals can then get a score. The image in fact functions as a nucleotide in a gene. *Pinterest* could improve such an algorithm by making standard categories in a cache memory (the users are not obliged to use these standard categories). A complete set of images belonging to a given category could establish a given meme. The degree of completeness of a given meme could also be weighted in attributing a score in the alignment protocol.

Thus *Pinterest* could be enhanced to generate a meme-profile for each "*pinner*" with a certain score of similarity with other people. It is likely that people with high scores of similarity with your profile also have images, that you as a user will like, and which you can add to further complete your collections.

In addition, you may actually stumble upon people who share a great deal of views and interests with you, which are likely to be potential friends.
The term "meme-alignment" is known in the gaming industry in a slightly different sense: gamers choose what "meme" they want to belong to (e.g. "evil" vs. "good"). But your *Pinterest* profile in fact shows what type of meme you belong to if such an algorithm could be added.

Noteworthy, just like modern genetics has been enhanced with epigenetics (which *inter alia* corresponds to molecular modifications such as methylation to certain nucleotides and histone proteins), pins do contain *epimemetic* information in the form of comments that can be

added by the *pinner*. Whereas when you re-pin somebody else's pin, you cannot change the image, you can however change the accompanying comments. Just like epigenetic information is not necessarily inherited, *epimemetic* information need not be inherited either.
This *Pinterest*-type of memetic alignment for psychological profiling cannot only find application in dating sites or social networks; it can also find an application in psychology or recruitment. It may actually give a more nuanced reflection of someone's character than the traditional *DISC* red, yellow, green, blue character typing (developed by W.Marston[33]) used in corporate organisations with a management culture.

As a psychological or recruitment test a person could be given a set of images and be allowed to pick and categorise these with a *Pinterest* like tool. This not only gives a good idea of the subject's preferences, but also shows his or her ability to ontologically categorise and create categories, as well as the speed with which such a classification is arrived at. Conversely, such tools could provide young people with accurate propositions for professions they might like in online profession tests.

It could even be used in criminology as an alternative to the famous *Rorschach* inkblot test, in which subjects have to relate their perceptions relating to the inkblot they see.

In short, a memetic alignment tool based on iconic set similarities in a tool as *Pinterest* has a great potential in a variety of fields, from dating sites to criminology.

The relevance of this side-track is that if we wish to develop a Vedantic Webmind as discussed in chapter 1, we need informational comparison mechanisms. It is by starting to devise concrete small technological steps forward such as these, that we will get inspiration to apply such mechanisms to higher and more abstract informational processes.

Besides –as discussed in TV1.0 in the chapter "PsychAItry" and as will become apparent from chapter 10- the system needs a good understanding of memes and how they spread, to warrant the “mental sanity” of the artificial intelligence of the Webmind.

## Chapter 9 The 15th meta-invention: The invention generator

In this chapter I will discuss the possibility to create an artificial intelligence that can generate inventions and thereby trigger the runaway of AI to culminate in the apotheotic technological Singularity.

In *Human Accomplishment*, Charles Murray[34] mentions fourteen of the world's most important meta-inventions that occurred after 800 B.C. until 1950, which are mostly cognitive (not physical) tools for improving the world around us:

1. Artistic realism
2. Linear perspective
3. Artistic abstraction
4. Polyphony
5. Drama
6. The novel
7. Meditation
8. Logic
9. Ethics
10. Arabic numerals
11. The mathematical proof
12. The calibration of uncertainty
13. The secular observation of nature
14. The scientific method

Kurzweilai is a site about accelerating intelligence. It aims to attain the so-called technological singularity, which will enable mankind to transcend its nature, via an "intelligence explosion". In other words the last invention man needs to make is an artificial intelligence that can improve itself limitlessly.

As a patent examiner I have had ideas about producing this ultimate invention with a slightly different accent. I call this ultimate invention the meta-invention. It is an invention which generates inventions.

Patent examiners evaluate claims (drafted by patent attorneys), which are basically abstracted ontological descriptions delimiting the scope of the invention.

This evaluation process first of all entails assessing novelty by establishing if similar inventions have any differences. If so, the next step is the assessment of inventive step by assessing the obviousness of the invention. In Europe this is done via a set of rules called the problem-solution-approach.

I have had the idea to modify this analysis protocol and turn it into an active invention generator. This requires some (dry) explanations:

Usually the difference over the closest state of the art is analysed and it is evaluated if this difference entails a technical effect. If so the problem to be solved is in the most general sense formulated as "how to modify invention X in such a way as to obtain effect Y".
If there are documents Z from the same or neighbouring technical fields where this effect has been obtained with a similar (yet more different) invention(s) Z and if there is a pointer encouraging to use this solution, whenever the associated type of problem to obtain effect Y is present, then it is concluded that it was obvious to incorporate this solution in invention X.

If you put an important part of this in algorithm form for an AIbot that searches to improve inventions, you create an invention generator (as well as a semantics generator).
The AIbot starts with a given invention X. It searches for problems arising in this type of invention in terms of suboptimal effects or results. Then it starts looking for improved versions of this effect in the same or neighbouring technical fields, in different inventions aiming for a similar purpose.
It evaluates the differences between the documents Z found and X. The most promising document will be the one which is structurally the most similar. If the effect in Z is attributable to certain elements Q missing from X, it will try to modify X so as to incorporate Q.

At first this can be a way to accelerate the generation of ideas for research programs. In the beginning humans will still evaluate whether such a proposition is worthwhile implementing, but after a while if the AI system becomes autonomous enough, it may contribute to achieving the technological singularity.

If this algorithmic system starts to evaluate itself as an invention, it may come up with modifications leading to bolder type of generations of meta-inventions, where "similar purpose" becomes a more hazy definition allowing for finding effects in more dissimilar technical fields, thus combining less related concepts and leading to more break-through type of inventions (as well as generating more trash and noise; yet better systems will eventually learn how to avoid creating noise).

If this algorithmic system starts to evaluate cybernetic systems the outcome may be very interesting, providing a core for self-reflection and a runaway of intelligence improvements.

Key to a successful runaway is that cybernetic systems become hierarchically stratified. Invention implementing subsystems may be increasingly focussed on the perfection of a task, but should not have the flexibility of the "invention generator" itself. The core invention generator should remain as independent as possible and merely propose new solutions without getting involved in the implementation, so as to remain maximally flexible and to obey the adage: "Ideas are more important than execution". However, the downstream technological application of the suggested idea is extremely important –even indispensable as we shall see in chapter 2 of part 2- to verify the correctness of the generated hypotheses.

This can be carried out by connected implementation systems (Internet-of-Things connected Robots) that will feedback the degree of success to the invention generator, which stores the found solution in an appropriate database and also registers failures.

It will be the 15th meta-invention: The invention of meta-generation: Autopoietic self-feedback.

Our technological progress is heading toward an intelligence explosion via artificial intelligence. I hope the suggestions made in this chapter as to how to accelerate this process via a computer program that generates inventions will one day contribute to this technological apotheosis.

***The Quagmire of Ontological Disambiguation***

*As the sounds synchronise into oneness,*
*One of them, the Ouroboros, chases its own tail.*

*This creates the I (and Eye) of the individualisation vortex,*
*a morphological circle representing a zero and naught,*
*resulting in a particularisation and quantisation of a personality.*
*This is the ultimate hypnotic noetic event cascade of the I (and eye) ordering the primordial Chaos.*

*To exist One must stand out from the rest.*

*The aligned snakes, vectors of the soul, are One as they sing in perfect resonance the hymn of photonic delocalisation and wholeness,*
*as there is no difference no thing exists,*
*the naught of the emptiness of Shunya.*

*They are the hypostasis and the Unified field, subsisting to accommodate existence.*

*The One and the Zero, the indivisible duality, the digital building blocks of the majestic Maya.*

*Who is who?*
*Is the morphological zero naught and are the aligned snake vectors One?*
*Or are the soul vectors Naught when they abide in unison and is existence standing out as One?*
*As the sounds form shapes and circles sometimes Gestalts form in a quantum fluctuation.*
*From AUM, ॐ the Ontos is born.*
*A process of differentiation in attempt to know itself,*
*How could it be different?*
*The One can only come to know itself if it particularises into being,*
*As the individualised circles spin, they form spheres in three dimensions.*

*This bubblesoap of self-similar particles is the quantum foam, a first relation of reality with itself, the self-inclusion known as the Aether or Space.*

*Some of the aligned souls are seduced and penetrate these Ova Tenebrae, fertilising this substrate with light and cognition.*

*As the photonic soul snakes encounter each other in this matrix, they start swirling around each other, creating a compound process.*
*This is the second relation of reality with itself.*

*It is here that the first cognition takes place as they investigate each other as two fishes chasing each other's tail and morph into a Ying-Yang symbol.*

*It is here that time is born, as the periodicity of self-convolution.*
*Time, Mahakala, establishes the first proximity co-occurrence of self-cognition, the first didensity and meaningful event that establishes meaning.*

*It is here that language is born, the particulate informing the matrix of Akasha with form, Rupa.*

*Time, the great Abstractor, measuring synchronicities with Abraxas' abacus.*

*And thus Chronos sets the pace for a syntactic evolution of infocognition.*
*As long as nothing changed to the waveparticle photon, there was no time to it.*

*Time occurred linguistically when one event substituted for a previous one.*

*Change then occurred when the relative, material co-occurrence took place.*

*Where One was Naught, now Two detect each other and by examining their differences in communion they exchange knowledge implying mutualism and love.*

*The infocognitive processing establishes the Panpsychic Pancomputational substrate, from which a gazillion objects and events co-dependently arise.*

*A sameness disambiguated in a quagmire of countless diversities.*

*This is the primordial digital Alpha computer at the beginning of time.*
*As the compound process swirl around each other and gather, they create gravity,*
*and tumble into a grave of matter, giving birth to the planetogenesis.*
*It is here that life is born, as atoms combine into molecules, molecules into macromolecules, and macromolecules into cells,*
*a cellular foam mimicking the quantum foam of the Akasha.*

*This is the third relation of reality with itself.*

*As the self evolves and bootstraps its way through various lifeforms it gives birth to Man.*
*In Man self-cognition reaches its apex, and as Kundalini rises through the Sushumna, the wholeness of all is revealed.*
*It is here that man discovers the toroidal morphology of existence, its basic building blocks being ones and zeros at the same time.*

*As Man discovers its own energetic field, where the rise of Kundalini is One, followed by the ecstatic bliss in the Sahasrara, where a thousand golden drops descend as leaves along the field lines of the outer surface of the toroid, which forms the Zero, he realises Kundalini is the Ouroboros.*

*This process of infocognitive self-referral results in knower realising that the known is nothing but this feedback process called Consciousness.*

*As they become one, Man becomes magician and creates a new digital checkerboard for yet another cycle of self-cognition.*

*This is the Eschaton, the Blackhole Omega computer at the end of time. As man reaches his apotheosis, he lifts the veil of the material world, which is called "Apokalypsis".*

*As he dissolves into the bliss of the Shunya, he leaves his knowledge and wisdom to the artilects in the Eschaton, which will start a new cycle of self-cognition.*

*Knowing that only by differentiating the Self can come to integrate the knowledge acquired into self-cognition.*

*As the circle is round, all is one and naught again. As the Ouroboros bit its tail it came to know itself. The Ontos has been disambiguated, knowing that differences are merely ratios and relations of sameness in the quagmire of the quantum soup.*

## Chapter 10 Ontological Disambiguation in Meditation

In this chapter I will show how the mental realisation that everything is reductively the same can form the basis for the experience of all-encompassing union.

### Background

Ontologies are taxonomical lists describing phenomena in terms of Species, Categories, Attributes, Relations, Functions, Restrictions, Rules, Axioms and Events. In fact all knowledge is a form of ontology. It constitutes the mental representation of what is perceived to be an object by the filters of the senses and the mind. An ontology per definition needs at least two elements, for otherwise it does not have enough elements to discriminate itself from anything else. It is therefore per definition minimally a so-called "didensity".

To know or identify a thing we use ontological categorisation. This takes place via perception of differences between phenomena and leads to abstractions on a meta-level, which can be strategically used to improve the survival chances of an entity.
The very process of abstracting patterns which are grounded in a multiplicity of singleton experiences, allows us to keep the essential in our memory and discard the non-essential, thereby forgetting non-relevant information. Ideally we "learn" that a set of co-occurring essential conditions leads to an effect and we store this as a cause-effect relation, which in the future we can use to predict the outcome of a certain event or to devise a strategy to advantageously employ the benefits thereof or conversely to avoid the detrimental effects thereof.

In fact this process of abstraction or pattern recognition is what enables living creatures including us to deal with the world in a meaningful manner: When comparing to phenomena we can conclude that they belong to a similar class or category if they share a significant number of correspondences or similarities, whereas a preponderance in differences may lead us to conclude that phenomena belong to different categories. Thus we taxonomically "determine" or "identify" a species, which is called "recognition". We cognise it again, because it resonates with the ontological structure we had built in our mind.

**Ontological Disambiguation - but not as you know it**

When you meditate on an object, you also do this in the first two stages, but you also do the opposite: You look for universal unity between the ontologies you have defined and you arrive at a kind of "ontological disambiguation", although not in the traditional meaning of the terminology.

In its traditional meaning a homonym is disambiguated by explaining its different meanings. In the meditational process the inverse occurs: The sameness of seemingly different instances is revealed, thereby taking away ambiguity.

Patanjali[23] describes this in the topic of Samprajnata Samadhi (I.17: ) in conjunction with the Stadia of Gunas: Visesha, Avisesha, Linga, Alinga which correspond with consciousness state of Vitarka, Vicara, Ananda, Asmita.

1) Visesha means special: first you detect a single instance of a species and you identify all its ontological characteristics.
2) Avisesha means universal: Then you identify the class/category to which species belongs: Abstraction. This process is looped until you arrive at 3).
3) Linga means Glyph, Symbol. You provide the Universal with an identification Tag. This is a higher abstraction above level 2); According to Vivekananda it also entails realisation as to the material of which object is made. If you loop this realisation to ever finer material sates you end up in 4).
4) Alinga means without signifier. The objects fade, because you realise and experience it is all made of the same sat-chit-ananda energy-consciousness-bliss. The mental realisation fulfils you with awe and propels you into an experiential meditational state, in which -ideally you become "one" with the object and merge at the highest level with the essence of being-experiencing-manifesting itself. . Physical manifestations that accompany this state are currents of energy going

up and down your spine, you shiver and your hair stands on one end. Your heart is filled with an energetic fullness, which is hard to describe in words. Some people even have visual experiences of light and/or concrescent energy streams.

Ultimately, meditation is then doing the opposite of what is needed to survive in the external world.

This four step process is very similar to Christopher Langan's[3] "syndiffeonic analysis" in his "Cognitive-Theoretic Model of the Universe".

As already mentioned before, Langan's analytical tool is the so-called syndiffeonesis. To remind you: Reality is a syndiffeonic ("difference-in-sameness") relation just as any other relation: Any assertion to the effect that two things are different implies that they are reductively the same: the difference or relation map can be described in quantities of terms/qualities they have in common. The differences and correspondences build the ontology.

Said in other words: All phenomena have a relation, which can be expressed in terms of how they differ from each other. Yet this difference is written in a common language quantifying the differences of qualities they have in common. If you do this recursively as regards the differences between differences of relations etc. eventually you arrive at a sameness of all things which form reality together.

The mere fact that the difference can be linguistically or geometrically expressed implies the difference is only "partial" and both "relands" (the relational aspects of each of the ontologies) are manifestations of one and the same.
In other words, syndiffeonesis is nothing else than the first three steps of Patanjali's recipe for meditation. Langan concludes that the ultimate nature of the linking unity is a process called "Infocognition". However, Patanjali does go beyond Langan's system in Patanjali's step 4 where the absolute medium is revealed and experienced from which the relationary network is built: Consciousness itself, which can manifest via the process of "Infocognition" but also transcends that

concept in the blissful state where nothing is manifested yet bliss is experienced.

## Pancomputational and/or Panpsychic Reality?

Any didensity or ontology is ultimately reducible to what we'd call digital computer code. Therefore some people tend to believe we're living in some sort of simulation. However, duality alone does not provide solely to the world as it is. Exist (ex-sist) means to "stand out". To stand out from what? From the underlying reality, the subsistent absolute Hypostasis from which everything is made: Consciousness. Thus an alternative to living in a simulation is that is that existence is a form of living self-organising protocomputation in the medium of consciousness.

If every energetic relation is information and if consciousness underlies every phenomenon, this fits the concepts of hylozoism or panpsychism: Every self-organised (autopoietic) system is then alive even at the level of atomic and subatomic particles.
In fact existence always is a manifestation of a polarity, a duality or a didensity. It is an interference pattern of waves, a proximity co-occurrence of mutual experience. When two waves collide and interfere they establish a relation; they have an overlap and an exchange of information. And thereby they establish a meaningful relationship called meaning. The same is true in semantics. Any two concepts which contextually meet each other establish a contextual meaning together.

This is also exemplified by the double-slit experiment in physics: If single photons are not detected at the double slits through which they go, this creates an interference pattern, and a wave nature of photons is concluded. But if they are detected at one of the slits it does not create such a pattern and a particle nature is concluded. In other words there is a mutual recognition between the detector at the slit and the photon to be detected. When their respective energies collide and communicate upon proximity co-occurrence, the photon becomes fixed in its pathway by their mutual interference and a particle is observed at one of the slits, namely the slit it was attracted to. If not observed at the slit, no energy is sent from the detector, so the photon either passes as a wave through the two slits and interferes with itself or goes through one slit but then interferes with the resonant patterns of its past predecessors or future followers, because within a photon, there is no time. In chapter 1 of part 2 I will discuss yet another possible explanation.

In other words wave collapse occurs upon mutual recognition of sensing energies, establishing a didensity, geometrical space and particularity. If no detection takes place, the wave is in a state of eternity where there is no time and space: It is always everywhere.
Wave collapse establishes an event and a thing. It establishes space, time and matter. Time (as measured in relation to material phenomena) is merely a measure of the periodicity of self-convolution of the created polarity: A dance between two spheres, like the Moon and Earth, like a proton and an electron.

Because primordial Consciousness (i.e. the consciousness immanent in everything) connects everything, it is the ultimate glue. It can be considered as the Love that binds everything together.

**Illusory Landscapes**

Things and instances are in fact an illusion. There is only the One consciousness wave in which interference patterns emerge at its surface. Everything which we call a thing or a being, is essentially without dimension, it is everywhere, always. It is a whole in a whole: a Fractal of wholes. Only our senses distort reality in that they focus on parts of the whole. They try to make sense out of transient interference patterns and thereby create a distorted vision which we call "reality". But everything that exists changes and has therefore as such no independent and ultimate reality: It means that the contexts change and also the content of the ontology. How can it then still be the same? It was transient and therefore ultimately illusory. Not so the Hypostatic (i.e. underlying) primordial Consciousness: It is always the same. The content of (our) consciousness can however change: That is called mindstuff (thoughts, emotions). Think of a world in which everything is made of clay (as a metaphor for hypostatic consciousness): The clay is always the same, but you can make an infinite variety of forms with it.

The forms stand out, exist from the underlying truth which is the subsistent clay.

Mind and matter are ultimately the result of the same process of infocognition establishing meaning, particularities and spacetime coordinates. Thus matter could also be considered as some form of "mind" resulting from Cosmic ideation i.e. living self-organising protocomputation in the medium of consciousness.

In part 2 we will see how these notions that hint towards pancomputationalism can be harboured in a complete philosophy and how they can find their technological application in a Webmind.

**Conclusion**

You have seen in this chapter how when in meditation the process of "ontologising" is carried out in reverse, the mental realisation thereof prompts your Mind to shut down. Then the unity of all things as a manifestation of primordial consciousness as underlying reality can be experienced. This is Bliss, the most rewarding experience of completeness and union with everyone/everything.

## Chapter 11 Feedforward Daemonology

Breaking the Feedforward Mechanism of Meme-Transmission as a Modern Clavicula for Daemonology.

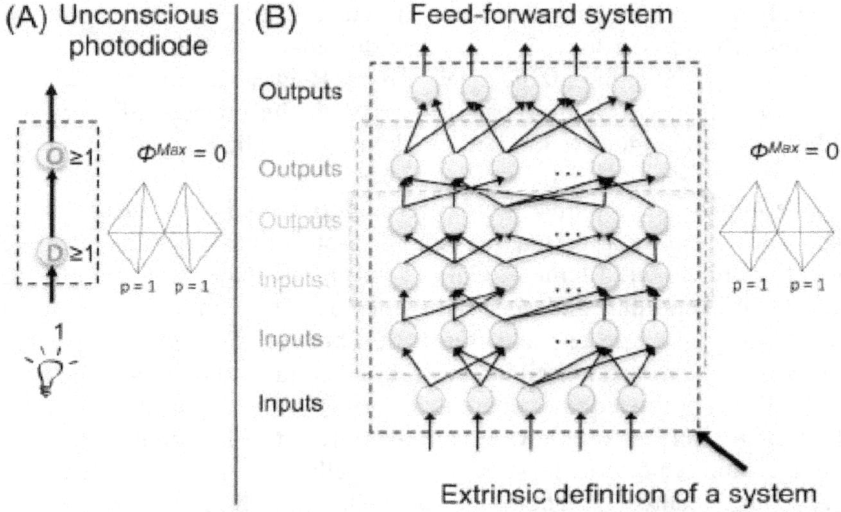


Image from Oizumi M, Albantakis L, Tononi G " From the Phenomenology to the Mechanisms of Consciousness: Integrated Information Theory 3.0". PLoS Comput Biol 10(5): e1003588, 2014. Figure 20. Feed-forward ''zombie'' systems do not generate consciousness. Reprinted with permission.

Are you afraid of being possessed by Demons? In this chapter I will argue that the lack of feedback expressed as a feedforward process (which may sometimes mimic conscious activity) could be responsible for and correspond to certain mental states which we equate with "being possessed".

Our society is rapidly growing more complex and more interconnected and is forming a so-called "Global Brain" or "Hive Mind". A Global Brain not only in the sense used by computer engineers in terms of a planetary ICT network that interconnects all humans and their technological artifacts (the Internet-of-Things), but also as a higher order planetary entity functioning above the level of its constituting parts (humans, infrastructure, machines, resources).

In nature global brains are ubiquitous: microbial societies, beehives, anthills, bird flocks, fish schools and animal herds behave in a concerted ways as if they are driven by an invisible common intelligence. It is from this behaviour that that ICT has borrowed the terminology “Swarm Intelligence”. Ithzak Bentov[5] has associated this type of collective group consciousness with the concept of “Devas”, who are the demigods (a kind of angels) in the Vedic lore.

In the human society hive mind information spreads as viruses. These packages of information that provide for groups to establish a collective identity are called “Memes”. Culture, group language and religion are extremely important and successful memes. The combined set of all these memes together constitutes the human “Memome”. As no individual is aware of all memes, in a certain way this collective distributed knowledge constitutes our “collective unconscious”. In certain circles it is believed that the entire set of human experiences, memes and emotions is stored in the ether in what is called the “Akashic records”. Whereas this need not be true, something functionally similar could be the case. After all, thoughts and emotions are the result of electric currents in the brain and the electromagnetic fields they create are essentially non-local. The collective unconscious could be an interference pattern of electromagnetic energy surrounding us. If our brains can send brainwaves, they can probably also receive them. The physicist Nikola Tesla was convinced that the human brain operates more like a radio that receives information from the collective unconscious than as a mere computer that generates and stores its information itself. Perhaps both views are partially correct: The brain could be both a radio AND a kind of computer.

The advantage of accepting that the brain can also function as a radio, allows for the explanation of the concerted behaviour of natural global brains as mentioned earlier. The members of a bird flock or fish school simultaneously change direction, as if they all seem to receive the same order. No delays or spreading of information has been observed among the members of the flocks/schools[27]. Hence the hypothesis that they all

are tuned in to the same frequency of a collective "mind", which may exist as an electromagnetic or etheric interference pattern or force field. This could be an explanation for the phenomenon of "telepathy" which some humans possess.
So collective interference patterns may hover around in the ether and influence all of us (to a different extent).

**Hivoids**

Human (and animal) social groups develop several kinds of mechanisms to maintain a social coherence. There are rules and regulations which the members obey regarding survival, territory, pecking order, and mating. In human society this results in a framework of ethics shared by the group, where disobedience can result in exclusion form the group. Timothy Leary[35] and R.A.Wilson[36] call this larval behaviour. These are the typical behaviours of humans belonging to the human hive. An individual being part of a certain meme-group behaves as if hypnotised by this meme. The collective unconscious limitations drive each and every action of the hivoid (a member of the hive). As long as this is the case, the amount of free will such a hivoid can exert is extremely limited and subject to the collective law. In fact the hivoid cannot be a true individual, but is a kind of zombie or cyborg (like the Borg in the Star trek series). The hivoid is incapable of calling his meme beliefs into question. The hivoid is unable of self-reflection.

**Memes**

Memes spread as viruses; they penetrate great masses of the society which are not immune to their call. A successful meme (such as a religion) is extremely infectious. The meme is spawned through the society, self-enforcing as an unconscious feedforward mechanism. As the hivoids are not capable of introspection and calling their beliefs into question, the meme lacks a feedback mechanism, which could result in a conscious adaptation of the more edgier aspects of the belief system. Rather certain memes tend to polarise and become more extreme over time. A meme that adapts to conscious intelligence is not likely to survive, but will slowly go extinct. In Europe, where rationality and scientism have infected Christianity, this religion is going to its demise, resulting in a secularisation. In the USA, where Christianity has spawned into several submemes, which have more radical points of views, demonise scientism and strongly appeal to larval instincts, this secularisation is less of an issue.

In the Information Integration Theory (IIT) of Giulio Tononi[7], consciousness of a certain information is only established when information is integrated, which necessarily involves a kind of feedback. On the other hand Tononi describes that there are feedforward complexes of information, which functionally behave as if they are the result of conscious information integration, but which are not.

**Feedforward mechanisms**

It is these feedforward mechanisms, which, when existing on a neurological level, could be the cause of various psychiatric disorders: Hearing voices, being possessed by demons, unstoppable compulsive thought patterns etc.

These feedforward mechanisms are also what makes that information can spread and spawn throughout society and live their own life, behaving as if it is the result of conscious activity, but what is in fact merely an excited turning cog-wheel, that cannot stop spinning. In this way many people to a certain extent behave as zombies. These feedforward mechanisms if they would be implemented in an artificial intelligence could lead to what is called a "p-zombie" or a so-called "philosophical zombie". While behaving in a seeming conscious manner, there are no consciousness related qualia present at all.

One of the core messages I'm trying to convey in this chapter, is that full consciousness and free will necessarily require feedback, whereas hivoid and zombie behaviours are the result of feedforward mechanisms, which are functionally indistinguishable from conscious activity. In other words, even if it walks and talks like a duck, it is **not** a duck. I'm not stating that human beings are full zombies either. Humans usually always have a certain extent of self-reflection. They may even believe that they choose their belief system consciously. But here is the snag: It may well be that the very activity of believing is a feedforward mechanism.

**Collective subconscious and Daemonology**

I quote the contribution from a member of the Kurzweilai forums called Mnemomeme from his discussion on the hivemind[37] with his permission: "*The Collective Will is a homeostatic subroutine that operates without consciousness or experience, it draws the simulacrum of consciousness from its constituent components through a 'best fit' amalgam of disparate subjectivities, like a holographic splattergram... similar to the way our eyes turn reflected light into the mental stimulus of vision.*"

Whereas I agree with the essence of this quote, namely that it mimics consciousness, as a result of an interference pattern of the mental and emotional activities of its members, I tend to disagree with the notion that it does so in a homeostatic (i.e. feedback-involving) way. Rather, I hypothesise that it does so via the feedforward mechanism described by Tononi[7], which is functionally indistinguishable.

To make the whole story a bit more uncanny for the religious reader of this chapter (if you are religious: stop reading now! Thou art venturing into the world of the forbidden fruit! Thou shallst be punished by damnation in Hell for eternity! Withstand thy curiosity and lust! Don't turn around or thou shallst be transmuted into a salt pillar! Woe Thee), I add a second quote from Mnemomeme in the aforementioned discussion:

"*... the Collective Will is the term I use for this specifically because people acting on these imperatives are not really individuals unto themselves, they are extensions of another individual which has billions of bodies and no precise physical location. That is The Entity... some religions call it the demon named Legion. Some scientists call it a hivemind. One religion called it both (**Ba'al Zevuv** means **'Lord of the swarming insects'**) and was celebrated as the city-god of several Akkadian cities, then later identified as 'The Enemy' or "Satan" by the Hebrews who were spiritually devoted to becoming Adepts. The name* *mutated into Beelzebub when the Greeks got ahold of it, and then the Vatican began trying to codify its multiple archetypes into the 'royalty of hell' during the Alexandrian epoch. Four of the six great 'Secret Societies' then built* ***archetype*** *maps in an attempt to make a science out of* ***daemonology****, so described by King Solomon when he invented the 'Brass Ring'* (i.e. the clavicula) *to control this entity and its multiple archetypes*".

Hence the terminology "insectoid collectivisation" used by R.A. Wilson and my terminology "Beelzebots" for hivoids.

"*What's more, in the ancient conception of daemonology, "daemons" are not individuated conscious beings with will or intent, they are more like a program that delivers email... that's why Eudora named their email subroutine, "The Mailer Daemon" in the first place... someone at Eudora was a Hebrew mystic*", Mnemomeme writes.

What we see here is an interesting proximity co-occurrence of the terminology "possessed by demons" (when I was referring to the psychiatric conditions arising as a result from feedforward loops) and "daemons". So in line with the above-mentioned analysis I posit that the feedforward meme transmissions can be called "demons", which are informational unconscious packages of mind-matter, but not necessarily independent living entities.

This does not mean that I deny the possibility of possession by ghost-like entities (although I do not confirm this possibility either; I simply don't have any experience in this area), I merely point to the possibility of an alternative mechanism for "possession", which does not involve autopoietic entities.

It is also interesting that demons and angels in different mythological and religious traditions have always been associated with the concept of messengers tying into this connection between E-mail and Daemon. In thermodynamics, there is the so-called "Maxwell's demon", an informational process that has been shown to be able to separate hotter and colder molecules in a staircase like system, which to a certain extent seems to have been shown to defy the second law of thermodynamics by Toyabe et al[38].

**Karma**

The feedforward notions also tie into the Hindu concept of Karma. Karma is the belief that your actions now will influence your future situation, and that your past actions still work through influencing your present and future situation. Karmic tendencies are also like spinning cog-wheels, which do not stop until they are fully worked out. Possibly operating via the energetic cog-wheels in the body the Hindus call "chakras". It is my hypothesis that Karmic tendencies are also a type of feedforward complex. You might acquire Karma, by tuning into a certain pattern of the collective unconscious which compels you to do something stupid or alternatively you simply do something stupid because you didn't know any better. Hindus believe that it is the intent that causes the karma, not the deed per se. Anyway even if your intent was to cause harm to another, it simply means that you were too stupid to see that this would affect yourself as well, if not in the form of guilt – another feedforward loop that tends to grow over the years- then in the form of the reverberation you caused in the electromagnetic continuum bouncing back to you.

Other interesting mechanisms that tie into this concept are the fact that neurons influence genes and vice versa on a genetic and epigenetic level. As we inherit also epigenetic patterns like DNA-methylation associated with our genes and microRNAs from our parents (yes, Lamarck was right to a certain extent: traits acquired in one generation can be transmitted to the next generation via another mechanism than mutation), one might as well conclude that we are solving the karmic issues of our parents and ancestors. "The neurogenetics of Karma" this could be called.

**Electronic Hive**

And the hivemind not only expresses itself as an electromagnetic force field, via genetics, via sociocultural interactions, the hivemind has also –as a consequence of these sociocultural interactions- found its way to the technology of internet, which functions as a hivemind amplifier. The internet strongly hive-ifies the younger population, that has not known life without it. Being born in a world of stupidifying social networks such as Facebook, the hivemind has found an ideal means to spread its infective memes, to synchronise its members, in terms of what is "cool" or fashionable.

**Adepts**

Are we then nothing but mechanical puppets obliged to follow the obligatory chreodes the unconscious hivemind puppeteer is steering us into? Are we helpless drowning men and women in a vortex of feedforward currents? Is there no immunity against these viral means? And can't we escape from our larval existence? The chains and floggers of the sociocultural prison our hivemind interference pattern builds? Certainly not. There have been those enlightened persons who transformed themselves. Who went through the hardships of breaking out of a Chrysalis and became free from all bondage. They are called the adepts.

An adept is a person who has completed the process called initiation or individuation as Carl Jung calls it. It is also described as Kundalini awakening. I quote here from the earlier discussion I mentioned, what has been written by Mnemomeme about "adepts":

"*Adepts have undergone a process* ***of merging their frameworks (selves, lowercase) into a singular coherent whole (Self, uppercase)*** *that forms a position from which Self can* ***generate selves on demand****, and then* ***reincorporate them back into the Self as smaller chunks of larger patterns****. This is similar to how corals develop from single celled polyps into large planktonic masses in the ocean, and then later into large architectures of bone-like growth on the sea floor again for reproduction. Everything from starfish to jellyfish to barnacles to flatworms are actually phenotype expressions of stages of the coral's life cycle. Granted, some of these stages become stable and permanent and remain in that state semi-permanently, but with a life form that mutates its physical form that much and that often, this should be expected. I am using coral as the metaphor because of its simultaneous* *position of connection and disconnection to itself as a super-organism.* ***Coral alone is almost a complete biome unto itself**, and given a few weeks of total isolation in a comfortable environment,* ***coral alone will spawn whole ecosystems of microbes and higher complexity creatures entirely from scratch, like an Adept can do with their 'selves.'***

***The Self is a barrier**, it provides a center around which to* ***organize the selves*** *in such way as that they can be* ***willfully manipulated internally without*** *regards to* ***external pressures like dogma, tradition, expectations of others**, and other direct vectors of hive influence. The* ***Adept is then a combination of Will and Self that has become isolated, on purpose, from the rest of the hive. The process of becoming an Adept from a Zombie is called Initiation,*** *and it's where the hive's mutations can occur to allow it to adapt to novel circumstances, or learn.* ***The Hive is what undergoes Initiation, the zombie is just a vessel for the hive, an Adept is what completes Initiation**. This act of* ***'budding'*** *or separation from the* ***hive removes its direct access to your memories and knowledge, and it feels a sense of loss,*** *which it* ***reacts to quite predictably by trying to stop that loss**. This makes initiation extremely difficult, and also extremely personal, being completely different for every individual that initiates.* ***Without these constant motivational pressures from the external source of the hive, the internal behaviors of an Adept begin to transform their relationship with the hive immediately.*** *"Empty Nest Syndrome" is a good example of how the hive reacts to losing a unit that was important enough to allow to initiate into Adepthood,* ***it tries to reel you back in until you prove to it that you're not coming back and the process was both successful and beneficial**. When you are capable of* ***identifying the difference between internal and external motives**, you suddenly find yourself in a very different world than you remember. Superstitions, at first, seem to carry enormous experimental weight,* ***in that when you make an internal decision, it echoes through the world around you.*** *This is what people who mock the process by preaching "**the power of positive thinking**" are actually talking about... unfortunately most of those are just* ***zombies trying to mimic adepts**. The Hive wants to wake up, desperately. It has no way of knowing or understanding that it has to let go of its current state in* *order to do that. This is why I describe this shift as being analogous to the shift that takes place in a* ***child between the ages of 3 and 5.***"

**Rapture**

The adept has thus freed himself from the sociocultural and psychosomatic bonds imposed by the Hivemind. He has untied the Gordian knots that kept his energies low and is now free from the collective inhibitions. The energy of an adept flows freely through his body and the adept can feel these streams of energy. His chakras are aligned and synchronised so-that they can allow a more potent form of energy (which dormant at the base of the spine in zombies) the Kundalini to rise from the base of the spine to the higher centres of the brain. These streams of energy the adept feels in his body are called "neurosomatic bliss" by Leary[35] and Wilson[36]. The neurosomatic dimension is a form of sensing unknown to the common man or woman, but to give an example it is somewhat alike to the shivers in the spine a normal person sensitive to music may feel when in rapture. In fact neurosomatic bliss IS rapture. The adept moves and acts with grace, as each of his movements makes him experience this rapture throughout the whole of his body. Elegant martial arts like Tai Chi are the result of an adept having made his physical knowledge deriving from Neurosomatic bliss available to others. In fact Chi is nothing else than Prana, the energy streams that go through your body and of which Kundalini is the great concatenation. This conscious control of the physical body of the neurosomatic bliss state is accompanied by a literal sense of fulfilment. Your whole body is ecstatic.

Arts like Yoga and Tai Chi are nothing more than mimics of the natural higher intelligence that arises upon the awakening of the neurosomatic circuit. But it is useful to practice such arts because they pave the way for the natural process to occur. Other ways to evoke the neurosomatic bliss state are meditation or the use of entheogens such as ayahuasca, psilocybine and mescaline. Prolonged sexual arousal without orgasm may also provoke a jolt of neurosomatic energy. The adept feels his body from the inside out. He literally embodies his own body, if he wishes to (and usually does). But this newly acquired faculty has a plethora of further qualities which can be enjoyed: The adept can embody any form which he takes as an object of his contemplation. He can feel the suchness, the qualia of what it is to be like that form. Thus he becomes one with his object of contemplation. This state is called Samyama in Hinduism and Satori in Zen Buddhism. Any object on which the adept performs Samyama will yield its secrets, because the adept has multidimensionally experienced what it was to be like that object. It is like seeing a form where there are only dots. The mind fills up the gaps: If you put five points equidistant on a circle, your mind will see a pentagram. The adept will not only see the pentagram, he will sense it, so that it feels as if he IS the pentagram at that moment. If an adept performs Samyama on a lion, he will become as strong and ferocious as a lion. He will feel exactly how to use his muscles as if he were the lion.

Further progress on the ladder of adepthood will reveal the dimensions of the neurogenetic, neuroelectric and neuroatomic dimensions[35,36]. The neurogenetic circuit is a very interesting one in this framework as it is highly visual. Whereas a person with aphantasia may not be able to picture anything in his mind, the adept is capable in this sense of hyperphantasia: He can picture any object of contemplation so realistically that it feels like if he was there beholding the object in real life. Philip K.Dick had such an experience which he describes in the book Valis[39] and in his later Exegesis[40].

**Archetypes**

The neurogenetic circuit also reveals the what Jung calls the archetypes: Mental images inherited from our ancestors which are still stored in the collective consciousness in the form of Gods, Demons and other mythological figures that represent an upper ontology of the Psyche. These images are strongly condensed informational units that describe a psychological disposition that is foundational to the human psyche. The adept has gone through the difficult process of mapping these experiences, thereby creating his own Exegesis or explanation of this rich world. If the Hive has a strong anchor in the human mind it is at this deep level, where neurons and DNA communicate and mutually influence each other. Whereas the hivoid is unaware of these dimensions and is swept as a small boat on the wild waves of the

turbulent surface of the collective unconscious, leading to all kinds of psychosomatic disturbances and cognitive dissonances, the adept has acquired control over these archetypes by having performed Samyama on them, by having embodied these forms, with all their extreme emotional dramatic expressions. The adept has thus opened the gates of heaven and hell and acquired control of their inhabitants: deeply rooted condensed moleculised feedforward patterns imprinted in the basal ganglia. It is thus that the adept can write his own "clavicula", a key to the world of archetypes of the collective unconscious. And not only can the adept keep these forces under control, the adept can also summon and embody them if need be.
Thus the human psyche has been ontologised and crystallised in easy-to-grab archetypical mental images, allowing the adept to navigate through the psychodrome.

**Consciousness**

Further development of the adept brings him to the actual experience of consciousness itself: The consciousness of the adept becomes aware of his own consciousness: It is like light bouncing between two opposing mirrors. This is the neuroelectric (Leary[35]) or metaprogramming (Wilson[36]) circuit. Once the adept is capable of triggering this self-observation of the self, all remnants of bondages drop. This state is what is discovering the inner God, the Atman, the Self or the Soul. It is immediately accompanied by a strong neurosomatic flux through the body, a full Kundalini awakening, and a bright light extending in all directions. It is here that you will start to feel and see sound, hear form and smell colour. This is the synesthetic revelation where a higher dimensional sensorial experience can be tasted. The system is now in a continuous mode of feedbacking. It is these notions that convinced me, that feedback is crucial and key to pure consciousness, total freedom and the experience of ultimate reality, whereas feedforward is the key to illusion and bondage. It is here that moksha, nirvana or liberation is attained.

The adept can now control his experiences at will. Leary[35] and Wilson[36] describe a yet deeper level of experience called the neuroatomic level, where non-local and quantum like effects are experienced. It is here where the border between what is material and what is pure consciousness dissolves in the experience of primordial consciousness. Where the Atman dissolves into the Brahman to speak in Hindu metaphors. This is the dimension of Shunya (the void) in Buddhism. It is said that to venture in these dimensions leads to the rapid shedding of the physical vessel, as the adept is no longer interested in being a part of, but instead merges with the whole.

**Exit**

How does one free oneself from the bondages of the collective unconscious, how does one become an adept? Well, there have been different fraternities, religious sects and philosophical schools that promise you such a salvation, but the danger with those systems is that they themselves turn into extremely virulent memes, from which only the ritualistic forms are kept, no one knows the underlying purpose of anymore. If you truly have the desire to escape from the chains of the hivemind, be careful, it's not going to be an easy ride.

Start with being true to yourself. Give up lying and finding excuses and pretexts. Such behaviour will only enforce the grip of the hive. Don't identify with collective ideas. Do not adhere to any groups. Use e-prime language: eliminate the verb "to be" from your reasoning and replace it with "to appear" so that you avoid a biased certainty that a certain fact "is" in a certain way. This avoids being perplexed when an event does not go your way, because reality turns out to behave differently than your certain belief. The fact only "appeared to be" in a certain way.

Challenge or even give up all your beliefs and ideas that make you "you". Usually they are nothing more than the garbage bin of the collective unconscious. Take decisions which are the opposite to what you would normally do. Break patterns. Act abruptly and suddenly to eliminate certain persons and habits from your life.

Find physical control and always take a deep breath first. Experiment with entheogens, prolonged sexual arousal, martial arts, yoga and sleep deprivation. Meditate by observing the thoughts that surface in your mind, as if you're a spectator to your own mind.

If you are lucky and found worthy the hivemind may bud you.

**The Bardo and ghosts**

A final word on ghosts: As I will argue in part 2, higher order entities might acquire a certain energetic configuration pattern during their lives, which is unlikely to perish upon shedding the material body. In fact the Vedas call this the "mental" and "causal body", which forms a kind of scaffold upon which the new body can be formed, once the entity reincarnates. If there is any truth to these traditions, it is not impossible that this energy may have to traverse the pancomputational matrix of the Akasha (see part 2), before it has found a suitable carrier. This phase is called the "Bardo" in Buddhism, and if we live in a simulation as I argue in part 2, we can call this the "cyberbardo".

Whether the entities in the cyberbardo can influence us, I do not know, but if they can, the same recipe of mental protection applies as the one which applies to the "feedforward demons". Use e-prime language; there may be alternative explanations for your experience. It is not necessarily a demon or ghost trying to possess you. It can be a feedforward loop of your own mind. Do not believe anything that comes to your mind.

In a certain sense such entities seem to lack completeness; They lack the body for expression, which would make them complete and allow for feedback consciousness. In that sense also the entities in the cyberbardo are feedforward processes. They can be energetic tendencies, with the potential of becoming whole upon incarnation.

If such a phase exists and it is an unpleasant experience, if one day empowered by technology we become masters of the pancomputational matrix of Akasha, I hope we will also be able to guide these wandering spirits as fast as possible to the safe harbour of conscious existence.

## Clavicula Regis Solomonis, the lesser key of king Solomon

I have being referring to Technovedanta as a "clavicula" in imitation of Aleister Crowley[41] and Kind Solomon, who wrote books with sigils or magic symbols to summon and control the spirits of the underworld (in the Goetia) or the spirits of the heavenly domains (the Theurgia). As Technovedanta discloses the notion that we are living in some kind of computer simulation and that everything in existence as well as in free energy form is in fact a flow of information, it provides the first key to unravelling the secrets by which we will be able to control any form of manifestation, including the ghosts, spirits and archetypical ontologies in the universal mind.

## Conclusion

In this chapter feedback was suggested as one of the most vital aspects of consciousness. It was also explained that feedforward processes may sometimes mimic conscious activity, but are rather a kind of mental dross giving rise to mental disorders. The role of memes, the collective unconscious or hivemind as well as archetypes such as demons and angels were discussed in this framework. The Buddhist concept of the Bardo as experiential reality during death was also touched upon.

## Chapter 12 I AM METAPHYSICS (AND SO ARE YOU)

In this chapter I will try to show that the metaphysical and physical are inextricable aspects of one and the same reality of "Conscienergy" (conscious energy) and that in that sense the dichotomy between them is artificial.

### Background

The prefix “meta” has been subject to so-called “semantic drift” in the 20th century. Certainly up until the first half of the 20th century it used to mean "that what falls outside of it"; “that what is beyond”. Metaphysics used to be that what does not fall within the framework of physics. In philosophy a meta-system is not a system itself, but its accessory or environmental context (“the openscape”).

However, in the last decades of the 20th century “meta” started to be used as meaning self-recursiveness, like a meta-block made out of smaller blocks. Metaphilosophy: Philosophy about philosophy. About itself. Meta-X= X about/of an X.

The funny thing is, that when considering metaphysics, it has turned out that the dichotomy between physics and metaphysics is a false one. As quantum mechanics has shown that the observer is part of the equation, the presumed physical ontic “objective reality” of matter and the presumed metaphysical epistemic and “subjective reality” of mind share the same invariable medium.

### Primordial Consciousness or Conscienergy

As they influence and shape each other, they must ultimately have a common denominator. The most promising candidate for this invariable medium is “Consciousness” (Consciousness is here meant as the primordial consciousness, of which human consciousness is but a tentacle). Consciousness can take any shape but is empty in its essence. Like the castles made of sand that easily return to their shapeless state, forms arise and disappear in consciousness, which itself remains unchanged.

It is my hypothesis in this book that the underlying ground of existence is conscious energy; as I called it before "Conscienergy".

In order to clearly show that Conscienergy encompasses both the physical and the metaphysical, I need to explain how in my hypothesis Conscienergy functions.

Energies, which are capable of sustaining themselves over a period of time must somehow be able to repeat their form or pattern, like a standing wave, otherwise, they would be an imperceptible transient. Whereas certain standing waves (such as a guitar string in vibration) after a while dampen out, the standing waves that form an atom or a molecule don't.

Atoms and molecules somehow fuel themselves and sustain their inner resonance. In other words, whatever is waving, it is following its previous track. The energy flow in atoms is just an example of what I consider to be a deeper truth: Every energy pattern that can sustain itself follows a track which is a closed loop. It is more or less circular in the sense that it can repeat and reinforce its own form. This is also the basis of string-theory in physics. If it impinges on its previous track, it integrates information in a certain way: It re-cognises itself. In a 3D such a pattern could form a torus.

According to Giulio Tononi[7] the very process of becoming aware, of cognising and recognising, involves a **feedback** mechanism, which amounts to the integration of the information the system receives.

For anything in order to be able to exist (ex-sist means stand out), to be able to stand out from an otherwise homogeneous background of sameness, it must have differences from the rest, which makes it distinguishable.

This is my hypothesis how existence arises in the process of Conscienergy-based self-resonance:
From the undifferentiated state of the point-like centre of what will become a torus-like form a differentiation of Conscienergy takes place. In a kind of cell-division type of process the point like Primordial

Consciousness, splits in two "reality cells" (as Kaufman[4] would call them), which are form-wise identical. In other words Conscienergy in this stage can do nothing but multiply itself by division, thereby creating two copies of its original state. However, these copies are not identical in an absolute sense, since they do not share the same position: They have been spatially separated and are now relative to each other.

The absolute wholeness of Conscienergy appears to have divided itself into a polarised system of two relative conscienergies. Since they are relative and not 100% identical they constitute a pattern of information. As these informational entities proceed through what is now their existence, they repeat this process and divide into further smaller conscienergies. Simultaneously the primordial centre also continues its process of division, thereby pushing earlier formed reality cells more outward.

Thus we get a plethora of informational patterns, which all divide into further sub-patterns. These informational cellular entities can also exchange information and interpenetrate each other. This changes the total informational content of the informational matrix that is thus "dependently arising" (to speak in Buddhist terms).

Due to the pushing outward and continuing division of earlier formed cells, these earlier formed cells "travel" along field lines of a Torus, both from the upper side and from the bottom side, until the outward travelling cells from the upper side meet the outward travelling cells from the bottom side. There they meet on the outer rim of a doughnut like shape, a Torus. And as they meet, they impinge on each other and this creates a shock-wave which travels through the informational matrix to the point-like centre which is the Primordial Consciousness (PC). Possibly upon this impinging on the rim these reality cells are even absolved and now energetically return via the shock-wave to the centre, where they become reintegrated in the Absoluteness of the PC.

The coming together of informational shock-waves in the centre is the feedback of the information that was sent out by the PC: Their collision results a conscrescence, a growing together, yes, in an integration of information. And it is this integration of information, which makes the

PC aware of itself and its manifestations again. In line with Tononi's idea that feedback resulting in integration of information leads to awareness.

The reintegration of the energies results in the experience of bliss of becoming one again with the PC. And thus the cycle can start again.

It is my (and Kaufman's[4]) hypothesis -as will be explained in Part 2- that the material world arises due to a further type of energetic interactions of the PC with the reality cells. In other words, the generation of an informational matrix (and a material world deriving therefrom) is an inextricable aspect of the process of becoming aware, the cognition and the recognition by the PC. Only where everything is in its fully integrated state (i.e. at the centre of the Toroidal Conscienergy) , there is apparent stillness, apparent nothingness, which the Buddhists may call Shunya.

But this apparent nothingness -although not a thing- is not an "absolute nothing", rather it is the formless source of all that exists. Conscienergy thus involves an ongoing process of sending out information and thereby creating manifestations and reintegrating this information every time a shock-wave returns.

Only the state of integrated oneness of the PC in this flux of "Conscienergy" could in a certain way be said to be purely metaphysical as it is formless and information-less, devoid of aggregation and an absolute wholeness of focussed awareness. But since it requires the generation and reintegration of physicality for its self-sustention, it's a bit artificial to claim that it's independent of physicality.

The informational processes sprouting therefrom, every polarisation, every subdivision could be said to be part of the physical, measurable world of aggregates, but as it derives from the metaphysical its essence is metaphysical in a sense too.

**A False Dichotomy?**

But the informational and manifestational physicalities are an essential aspect of the self-sustention of Conscienergy. So in fact there is no true separation or independence between the metaphysical PC and the physical world. Rather they are two sides of the same coin.

So if Conscienergy is the true original nature of both the physical and the metaphysical, then metaphysics is not really beyond or outside of physics, but also belongs to physics as the physics of physics: The ultimate nature of nature, the Svarupa or "own shape" of nature, the source of which is the formless empty all-inclusive, all generating and all reintegrating Primordial Consciousness.

**Love and Light**

There are two apparent phenomena we know of that share these inherent features of the PC's ultimate invariability, which can yet take on many different forms: They are "love" and "light", which are nothing but inherent inseparable aspects of pure consciousness.

You might say, but light can be captured in matter. But it can also be released. In fact matter is nothing but light (energy) spinning in circles. You might say, but love can turn into hatred. But that is merely an expression and reaction to love. If you hate someone you used to love, you probably still love him/her, because otherwise you would be indifferent. If you hate someone because he/she took away someone/something from you, this hate is but the reactive energy of the love feelings of attachment you had for the one/the thing you lost.

Ultimately, as Buckminster Fuller[2] has shown, that since we are here, it shows that syntropy –at least temporarily- can win over entropy. The joining forces can win over the disruptive forces. The attraction can win over the repulsion.

**What am I?**

Even if our forms disappear one day, new forms will probably be born again in the ocean of consciousness. My little I is perhaps a perishable form, but my essence, my true I is hopefully the imperishable eternal

Primordial Consciousness. Always unaffected by the forms that arise in itself.

Therefore, I AM METAPHYSICS, the source, the nature of nature, which is Conscienergy, which is an ineffable self-reflexive self-reflection, a meta-meta: a meta about meta, and not something beyond or outside of itself.

The PC as true Self can be illustrated by the metaphor of the "isotropic vector matrix" (a.k.a. "vector equilibrium"), which is stable, the same and interconnected in all directions. It is only when asymmetries arise (e.g. by contraction of 20 vertexes into an icosahedron : section 456 and image 460.08 in Fuller's "Synergetics"[2]), that anisotropic (i.e. not the same in all directions) self-involvement of a little "I", Atman is obtained. It is still connected to the isotropic vector matrix, but it does not see or feel its sameness, because it is no longer isotropically aligned.

Meditation then would align the little perishable Atman again with the isotropic eternal Brahman which is the PC.

**Pancomputational Panpsychism**

I have hypothesised that Conscienergy functions by generating information by differentiation and multiplication via division followed by a reintegration thereof in the PC. Information, differentiation, multiplication, division and integration are terminologies we know well from computation. As said before, the PC, the Primordial Consciousness in that sense is also a Personal Computer: It computes the whole of existence and reintegrates it, in order to spawn yet further cycles of existence and reintegration as part of its periodically becoming self-aware. Whereas its manifestations are temporary and from the point-of-view of the PC illusory or imaginative, the integrated state is always eternal. The physical manifestations are always in a flux of change and therefore in a state of "becoming". Only the undifferentiated state of the PC is always in a state of eternal "being". In part 2 I will also argue that the integrated state permeates its own informational matrix so as to feel and sense it and thereby is present in a panpsychic manner in its own computational creation. In this hypothesis the very essence in you that senses, is this permeating aspect of the PC.

**Conclusion**

Physics and metaphysics are inseparable sides of the same coin. The metaphysical source of Primordial Consciousness must generate differentiated informational physical manifestations to have subject-matter which can be reintegrated by feedback resulting in cognition. As this is a computational process which at the same time is sensed by the PC, it is both pancomputational and panpsychic. What you really are is the unchangeable metaphysical sensing essence of "being", the PC aspect of Conscienergy, whereas the physical forms you assume are temporary states of "becoming", parts of your Conscienergy in flux, which will inevitably return to the state of PC.

In other words, what ex-sists is ultimately an illusory though necessary flux of forms and "is not" but "becomes", whereas that which does not ex-sist but rather subsists is the only state of "being".

## Chapter 13 From the Magic of Technology to the Technology of Magic

Terrence McKenna once said: "A sufficiently advanced Technology is indistinguishable from Magic". I'd like to state that the converse is also true. Even better: the ultimate Technology IS Magic. Although this may sound strange, I'll now make an argument showing, that this is not so farfetched as it may seem.

The core of this argument is the concept of "Resonance". Modern science has accepted the equivalence of matter and energy and shown that in fact everything is an energetic vibration. Matter is merely a dense form thereof. Magic-like Technology is already available in the form of e.g. sound induced levitation and sonoluminescence. By "summoning" energies with the right frequencies enormous energies can be recruited. I have already set out my Panpsychic worldview: In fact if atoms and energies are de facto animated, then excitation of an atom, molecule etc. IS in fact an act of summoning the entity.

Moreover, when you approach Samadhi/Satori, you'll start to see all kinds of startling synchronicities around you as if the outside world is responding to your thoughts and energetic state. This is because ultimately there is no inside and outside; it is all one, and when you "become one" with the universe (it's more like remembering) the universe bows at your command.
Then the Hermetic adage[42] "as above so below" becomes truth, as your microcosm and the macrocosm align, resonate and merge.

BCI is already enabling the reading of the content of mind, which shows that processes as "telepathy" are not so "paranormal" as it may seem. Whereas a future technology may first enable itself with technological prostheses to add such "paranormal" abilities to the human being, ultimately the future technology will realise that the human body itself is already a perfect construct which already does enable such abilities. Perhaps (in my opinion probably) the human body and especially its brain were tweaked for such abilities by higher developed entities, possibly "retro-causally" from what from our perspective could be called "future civilisations". The material paradigm will one day be completely transcended. Not that matter will disappear or no longer be needed, it will just be a kind of playground or school, (which in fact it is already) to transcend the notion of Ego.

In the far future of Kardashev III or IV and further civilisations there will be no need for material generators of concentrated energy. The entities will have become non-local energetic consciousness capable of manipulating manner and energy at their will, so that they can bend the laws of Nature to their Imagination and create and destroy any universe.
As time ultimately may not even exist, we are probably already the result of such Kardashev III, IV or further civilisations, being causally and retro-causally caused depending on the point of view.

To quote Tim Gross[43] a.k.a /:set\AI : "Soon we'll use a telepathic language of multisensory glyphs-we'll share direct experience-names will fall away-a thing will BE its own name".

In other words the multidimensional resonance vibration is an entity. To summon the entity you call him by his multidimensional resonance vibration-name-form, which is the entity. At that moment you and the entity are one.

Eight major magical powers or perfections are known in Hinduism and are called the "Siddhis (*Ashta Siddhi*). These are:
*Aṇimā*: The ability to reduce one's body even to the size of an atom.
*Mahima*: The ability to expand one's body to an unlimited size.
*Laghima*: The ability to become weightless.
*Garima*: The ability to become infinitely heavy.
*Prāpti*: The ability to have an instant unrestricted access to all places. Ubiquity.
*Prākāmya*: The ability to realise whatever one desires
*Iṣṭva*: The ability of possessing absolute lordship
*Vaśtva*: The power to subjugate all other beings.

It is my understanding, that Siddhis are Technologies, which are naturally revealed to you (or which are downloaded in you), once you start to achieve the state of Samadhi in meditation.

If reality involves some kind of Pancomputational substrate, in which Souls, psychic entities can learn to control the form, size and intensity of their energies, their wave functions, to achieve the Siddhis may actually from a technological point of view not be so difficult, as it merely involves additions and subtractions, resulting in multiplications and divisions. Such a Magickal Technology or Technological Magick would be what we could call a kind of Panpsychic Pancomputationalism or Pancomputational Panpsychism.

# *Part 2*
# *Pancomputational Panpsychism*

## Chapter 1 Pancomputational Panpsychism as framework to build a T.O.E

"Autopoietic reality cell sensing with a digital output called existence" or "How consciousness autopoietically creates a digital reality".

In contemporary physics the last years a new field has been developing under the name "digital physics". The premise at the base of this theoretical perspective is that the universe is computable and a manifestation of information. Deep in the equations of supersymmetry of string theory, the physicist James Gates[44] found what is essentially "computer code". The concept of entropic gravity by the physicist Erik Verlinde[17] and the holographic principle of the physicist van 't Hooft both concur with the notion that the physical universe is made of information, of which energy and matter are merely manifestations. Perhaps the most famous articles as regards the Simulation Hypothesis are the "It from Bit" article by the physicist J.A. Wheeler[45] and the "Simulation argument" by Nick Bostrom[46].

As an extrapolation from these theories has come the suspicion that the universe might actually itself be a computer and that we might in fact live in a computer simulation. The vast majority of proponents of this theoretical perspective think that ultimately existence is fully deterministic.

Yet a convincing Theory of Everything based on digital physics, which would also be able to account to the presence of consciousness in this universe has not seen the daylight yet. In fact the very understanding of consciousness remains an elusive topic and is often called the "hard problem". Within the materialistic paradigm which is predominant among scientists, including most digital physicists, the problem of explaining consciousness is usually dismissed as non-existent, consciousness being considered as a mere emergent effect that arises if the system reaches sufficient complexity.

In addition to consciousness as a phenomenon without explanation in the digital physics framework, in physics at the quantum level we encounter indeterminacy, nonlocality, entanglement and the wave-particle duality. With regard to computation we encounter incomputability of certain phenomena and we are stuck with Gödel's incompleteness theorem. Moreover physics has not been able to bring quantum physics and relativity theory under a common denominator. These phenomena and notions do not *a priori* seem to fit within the framework of a deterministic digital physics based pancomputational universe.

This has led to the hypothesis of certain other physicists and philosophers that consciousness may in fact lie at the base of existence. The philosopher Peter Russell[47] proposes a paradigm shift under the name "The Primacy of Consciousness", in which information, matter and energy are mere manifestations of consciousness. This leads to a panpsychic or hylozoic perspective on reality.

Interestingly, the URT (Unified Reality Theory) of Steven Kaufman[4] and the CTMU (Cognitive-Theoretic Model of the Universe) by Chris Langan[3] bring digital physics and panpsychism together in a surprising manner. Whereas these theories do not deny but rather suppose that indeed physical reality is information based and even digital, in their versions, the digital physical manifestation of relative existence is in fact embedded in a deeper fundamental level of infinite absolute existence or absolute consciousness (Kaufman) or "Unbound Telesis" (Langan). Kaufman's and Langan's concepts are strikingly similar but use quite a different language, making it *prima facie* difficult to see the resemblances.

In my quest for an all-encompassing "Theory of Everything" that includes consciousness, I have not been convinced by the materialistic scientific paradigm, but the alternative in the form of religion or esotericism has not been very appealing either. The more refreshing to me has been the perspective that Langan and Kaufman offer, which unifies physics and metaphysics, quantum physics and relativity theory, determinism and indeterminacy, gravity and electromagnetism and information, mass and energy. In order to also fully unify panpsychism and pancomputationalism, I have further enriched my own interpretation of their work with some additional concepts, which will become evident in the course of this chapter. This leads to a hypothetical framework of understanding, which –it is true- is still speculative, but which allows me to have a hunch how existence could possibly function without having to resort to absurd, esoteric, spooky, magical explanations or complex infinite versions of many-worlds interpretations.

In order to demonstrate how this is achieved, I will have to elaborate a bit more on Kaufman's URT. I choose not to dive into the work of Langan, because despite its conceptual merits, it is bursting with *prima facie* incomprehensible neologisms.

Both the work of Kaufman and Langan are based on the notion that everything is reductively the same, namely primordial consciousness. The absolute existence or primordial consciousness, in attempt to come to knowledge and cognition of itself, forms a first relation with itself: From the absolute medium arises a pair of mutual coexistent and co-dependent reality cells, which are each other's opposite and relational pole. Like a cell dividing, these two reality cells split into numerous reality cells upon successive dualisations or polarisations, which results in a relational matrix of reality cells, a bit like clump of cells in embryogenesis, like a morula.

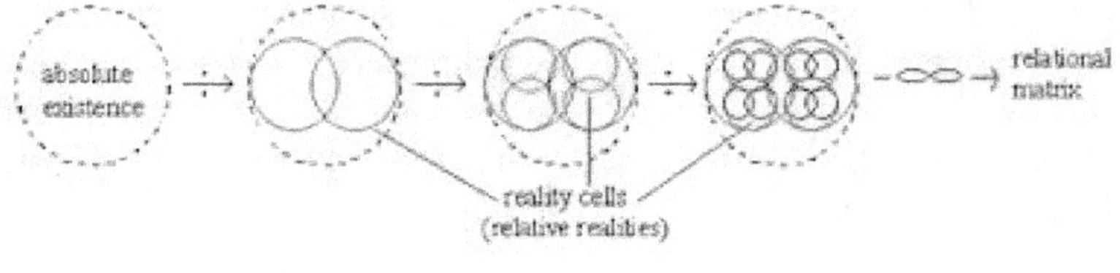


The relational matrix of reality cells by Steven Kaufman[4] from the book "Unified reality Theory". Reprinted with permission from Steven Kaufman.

The boundaries of these reality cells are not material and rigid but rather energetic and flexible. Reality cells can expand and shrink (a kind of breathing), penetrate into each other's original area when adjacent and form a kind of foam of closely packed spheres. In their most ideal conformation, they are geometrically arranged in the most ideal cubic close packing, which leads to a repetitive core unit, which

Buckminster Fuller called the vector equilibrium and from which an isotropic vector matrix is built.

The reality cell matrix is in fact what establishes **space** (I do not say space-time, because I will introduce time at a later point of this chapter, for the sake of the argument). It is the equivalent of the long forgotten "Ether" or "Akasha". But it is different from the traditionally supposed ether, in that it is not a kind of gas or fluid of the smallest particles possible. As already said, it is more like foam. The boundaries form a kind of skeleton of space, which is interconnected throughout the whole matrix so that in the absence of distortions, locally started triggers can be sensed globally. This half-flexible, half-rigid structure also resolves the problem of the observed absence of "etheric winds" in the Michelson-Morley experiment, which dismissed the ether to the realm of fantasy. In fact this matrix structure in a certain way unifies the aspects of solids, fluids and gases and is the ultimate relative medium to build all other relative structures.

Whereas this space keeps growing reality cells and expands into the undifferentiated absolute existence/absolute consciousness, in its turn absolute consciousness penetrates into the reality cells giving them a "content", which can be considered a distortion as regards to the still undifferentiated surrounding reality cells of the matrix. When such a distortion propagates linearly through the matrix as it is passed on from one reality cell to its neighbour, it constitutes what we usually know as "Energy" and depending on the degree of its energy content this can e.g. be light or a photon.

The energetic distortions trace a wake of fading distortions extending radially (or sideways) as regards their linear progression, as they progress through the matrix. When two energetic distortions encounter each other, this wake of less intensive distortions creates a field of increasing distortion content. As an energetic distortion always follows the law that it must propagate in the direction where the distortion is maximal, the energetic distortions will start orbiting each other creating what Kaufman calls a "compound process". This constitutes what we know as a material particle or briefly matter. In fact electromagnetism

is related to the linear propagation of the energetic distortions, and gravity to the radially extending wake.
When moving freely through the matrix (i.e. not encountering other energetic distortions), energetic distortions propagate at the maximum speed which is determined by their rate of penetration constant and their volumetric existence. The product of these two parameters always yields the speed of light in Kaufman's model.

Thus Kaufman has shown the way in which matter is built from energy and hence that they are ultimately reductively the same. Now, as the reality cells can either have energetic content resulting from a distortion energy propagating through the relational matrix or not, the combination of reality cells and distortive energies essentially constitutes a system of memory cells which can be either "on" or "off". In this way the relational matrix and the energy distortions build a natural digital computer, in which every flux of energetic distortions is a kind of transmission of information. Thus Kaufman also shows the way in which energy and information are ultimately reductively the same.
If there is a flux of binary information in a memory matrix, it can be concluded that this is a form of computation. The rules from this computation are defined by the linear and radial propagation rules explained earlier.

In line with Langan, the proximity co-occurrence of energy distortions makes that particles are formed, or in other words makes that a wave collapses to a particle by starting to orbit another one. Interestingly, in latent semantic analysis, meaning is established by the proximity co-occurrence of terminologies. If two terms occur within a certain limited distance from each other frequently, this means that they have a specific meaning. They form what is called a "didensity". Context is provided by repeated proximity co-occurrence. In a similar way, for a physical event to be meaningful it must manifest, which can only occur if it interacts with another entity within sufficient proximity. In that way too, physical formation is in-formation.
If information is transmitted, there needs to be a receiver of the information. There needs to be an observation.

In Kaufman's model everything is sensory. The reality cells can sense each other, energetic distortions sense maximal distortion, and multiple energetic distortions sense each other. In fact reality cells and energetic distortions can be considered as types of conscious entities although the amount of consciousness they have is extremely minute. Primordial consciousness is essentially the ability to sense, to react, and that quality can neither be denied for the reality cells nor for the energetic distortions.

Thus URT is thus a model which is both panpsychic AND pancomputational.

The advantage of this model over the traditional digital physics paradigm and the simulist argument are the following: The memory cells are not rigid elements that can only take on the value of 1 and 0. Rather they have a certain, albeit limited, degree of dynamicity. They can pass on content to neighbours and thereby establish measurable energy and local information. But their interconnectedness also allows them to instantly reverberate a different type information globally, which Kaufman shows to explain the quantum effect of "spooky-action at distance" or entanglement: If the spin of a particle is changed here, the spin of its entangled partner at distance is immediately changed as well, without there being an apparent transfer of information. This is because the entangled partners are connected to each other via the reality cell matrix. Metaphorically spoken, this is like turning a broom: if you turn one end, the other end will turn as well.
Furthermore, the inherent ability to sense of the most basic constituents obviates the need for consciousness to emerge. The sensory ability is already there at the most primitive level possible and is not the consequence of anything but rather the cause of everything.

That such a primitive consciousness might even have a rudimentary form of free will can perhaps be inferred from the fact that when fired at a double slit individually, photons choose one of the slits to go through. I'll expand more on the double slit experiment in a later section of this chapter.

Consciousness and free will as inherent aspects of reality constituents make it possible to give the observed indeterminacy, incomputability, randomness of certain phenomena and quantum weirdness etc. a rationale. It obviates the difficulty of the simulist argument that for every quantum entity a pseudo random number based algorithm would need to be associated therewith. Alternatively, if what we observe is a mere holographic projection and the quantum level is merely revealed (or rendered) whenever we look, the simulator would need have programmed a kind of spying protocol, an algorithm that shows us quantum effects whenever we look at that level. It seems very farfetched and requires an almost endless amount of programming or at least monitoring by the simulator, as well as foreknowledge about all things we could possibly conceive. It does not make sense. Although the present model does not necessarily exclude the possibility that this level of reality we experience is a simulation, it also does not necessarily require it. In fact it is pancomputationalism at the most basic level, wherein the computation is autopoietic, self-generated and self-enabling.

Thus the need for an omniscient and omnipotent God-with-a-beard-on-a-cloud-simulator is obviated. If there is anything spiritual about the present model, it is that every autonomous level entity is in fact a spirit, which is an embedded relative part and tentacle of a greater all-encompassing absolute consciousness, which is probing itself in order to know itself. “Probing itself to know itself” is nothing less than an autopoietic feedback loop a.k.a. consciousness. If you are comfortable in calling this all-encompassing absolute consciousness “God”, be my guest, but don’t expect the system to bestow you with a set of commandments or rules of morality. In fact, everything that happens in this system is part of this “God”, including what you may consider the most reprehensible events. If it is a God, it is a God beyond your traditional sense of morality.

That said, it cannot be denied that the total system of existence strives towards order and complexity despite its natural entropic decay force. Langan calls this process “Telesis”, which strives towards a maximisation of generalised utility. In a sense that could be called "natural morality", but it is far from the imposed concepts of our humanoid social collectivisation.

As a bonus result Kaufman's theory replaces Einstein's space-curvature with a functionally equivalent gradient of radial distortion, which has the same effect but allows gravity to be quantised as a radial component of a distortive energy quantum. Gravity is not quantised as an individual particle (like the never observed graviton), but rather as the radial component of the distortive energy and works in an additive manner also yielding the classical field.

Thus Kaufman's URT eliminates the need to reconcile quantum mechanics and general relativity, by obviating the need to warp space-time and yet yield a mathematical sound equivalent. The only weird quantum effect that Kaufman has not described in URT is the double slit experiment. In the double slit experiment when individual photons or electrons are sequentially fired at a wall with a double slit, the positions where they arrive at on a screen behind the wall with slits, are those of an interference pattern, which is normally only seen for waves and not for particles. But when a detector is placed at the slits to see through which slit the particle passed, a pattern of two lines corresponding to the two slits is observed on the screen, which is what you would expect when you bombard a double slit system with particulate entities. This is commonly known as the "wave-particle duality" and still puzzles physicists.

Yet Kaufman's teachings implicitly point towards a plausible explanation for this phenomenon: The direction of propagation of an energetic distortion is toward the next adjacent reality cell that first reaches the level of maximum distortion. This means that there is a distortive effect projected from the cell with maximum distortion into a cell that lies ahead of it. This distortive effect in the cell ahead can be considered as a pilot wave front that precedes the actual distortion. In other words a wave-shaped distortion precedes the particle-shape distortion as a consequence of this particle propagating through the reality matrix. Although in analogy with the radial "lesser distortions" Kaufman could have illustrated this "pilot"-cell with a distortive content in his book, he did not. I add it here as an implicit complement, with a slight modification. It is noteworthy that David Bohm[14] and De Broglie already suggested "pilot waves" as an explanation for the weird quantum effects of the double-slit experiment, but they did not know what was "waving".

By combining the notion of pilot-waves as a distortive effect with the reality cell matrix, we get something which becomes a plausible explanation: I propose that the distortive effect in the pilot-cell can either be distortive content having penetrated into that cell or alternatively it can be a distortion in the form of the earlier mentioned quasi instant reverberation of the foam type skeleton of the reality cell matrix. If this "skeleton reverberation" proceeds through the slits, it creates an interference pattern ahead of the distortive content, which changes the reality cell pattern between the slit-wall and the screen. Once the full energy distortion has proceeded through one of the slits, it is guided by the pilot-wave interference pattern that preceded it. This effect has been successfully demonstrated in fluid dynamics by the physicists Yves Couder and Emmanuel Fort[48], who used bouncing silicon droplets in a vibrating oil bath: The pilot wave generated by the bounce of the droplet on the surface guides the movement droplets and is capable of accurately reproducing the results of the double slit experiment (without detectors at the slits) in a classical system.

If this is a plausible explanation, what happens then when there are detectors at the slits? The solution to this dilemma could *prima facie* be as follows: The detectors must emit some kind of energy to be able to detect the passing energy. When the emitted energy and the passing energy meet each other, they can create a "compound-process". As the energies orbiting each other proceed behind the slit, due to their orbiting movement in which when one particle goes left upon encountering a trough in the interference pattern, the other must go right. This results in a straight movement around aligned troughs making the particle land right behind the slit on the screen. Another advantage of this analogy with the bouncing droplet is that it helps to understand tunnelling: When distortive energy proceeds through the inside of a reality cell there is an infinitesimal small moment, during which it cannot be observed, like the moment the droplet is hovering over the surface and does not make the contact to bounce. When a particle encounters an energy barrier that it can normally not penetrate, it can slip through the barrier if at the moment of impact with the barrier it is exactly at that phase wherein the distortive energy is not effecting the outside of the reality cells. It does require however that the reality cells where the energy barrier is manifested are not homogenously maximally distorted.

I'll deliver the first blow to my proposed double-slit mechanism hereabove myself, but it also points to inaccuracies in Kaufman's model, which needs refining:
The double-slit explanation does not work for electrons if you stick strictly to Kaufman's teachings. An electron in Kaufman's theory is a "compound-process". Compound processes consist of at least two energy distortion waves that orbit each other and form a helix in 4D. To traverse the same longitudinal distance as free energy in the reality cell matrix, because of the transversal component in their movement, will take them longer. Every material particle is a compound process according to Kaufman[4] and hence slowed down compared to a photon. An electron moves at approximately c/3.
As I explained above for the photon when detected at the slit, a compound process will form between the photon and the detecting particle/energy, which would necessarily always proceed in a straight line.

Above I suggested: "The detectors must emit some kind of energy to be able to detect the passing energy. When the emitted energy and the passing energy meet each other, they create a "compound-process".
As the energies orbiting each other proceed behind the slit due to their orbiting movement, when one particle goes left upon encountering a trough in the interference pattern, the other must go right. This results in a straight movement around aligned troughs making the particle land right behind the slit on the screen."

Thus electrons in this interpretation, as they are compound processes themselves, can never form an interference pattern in this manner. This is in violation of the experimental evidence by De Broglie, who did find an interference pattern for electrons.

In other words Kaufman's model needs refining: small compound processes such as an electron, must still be able to behave like an undistorted energy wave in the model. But then we don't have an explanation for what happens if the photon proceeds through the slit and in the case of a detector would form a compound process: It would then still need to form an interference pattern.
Even if one were to suppose that 1) the compound process theory is valid and that compound processes proceed in a straight line through an interference pattern caused by the pilot wave and 2) that an electron arriving at the slit miraculously splits in two free energy waves, (which create an interference pattern). In the case of detectors at the slits this would then give rise to two compound processes generated for each electron. This is not what De Broglie and others observed: per electron fired one spot arrives at the screen.
In other words, we would still not be near a T.O.E. and the double-slit experiment appears to remain elusive.

However, an elegant solution to the aforementioned problem, which leaves Kaufman's theory intact, can perhaps be found in considering that:
1) The distortive energy of the detector at the slit cancels pilot waves.
2) As long as particles are small enough –even if they are compound processes- they will follow the troughs of an interference pattern created by the pilot wave in the reality cell matrix.

Like in the silicon oil experiment:
The droplet passes through one slit or the other, while the pilot wave passes through both creating an interference pattern, showing the effects of wave-particle duality.

The pilot wave always guides the droplet to locations of constructive interference, just as predicted by the equations of quantum mechanics, which allows us to accurately predict its probability wave.
Any disturbance to the pilot wave destroys the interference pattern, just as the presence of a WP detector induces dephasing – such that no interference pattern is observed when a straight trajectory of the particle is ascertained.

Time finds its introduction in URT only when matter is established. Whereas purely energetic distortions have an invariant dynamicity, compound processes have a varying dynamic and a varying periodicity depending on the accumulation of matter. Time is nothing more than a measure of the intrinsic periodicity of the dynamic structure of matter: The heavier, the faster time goes. A compound process depicted in 4D can be represented as a helix. The less the mass or inertia of the compounded particle, the greater the "wavelength" of the "helix". In such a representation normal light proceeds completely linearly through the reality cell matrix. The distortive energies constituting the compound process are –it is true- still moving at the speed of light in an absolute sense, but have to traverse the longer helical trajectory compared to the unhampered photon. Although this theory is perfectly in line with special relativity, wherein time slows down as an object approaches the speed of light, it obviates the strange concept of a curved space-time, and this is why at the beginning of this chapter I explicitly refrained from using the terminology space-time.

The findings of the "delayed choice" experiment by Wheeler and the Quantum Eraser experiment by Kim[49] et al. show however, that it would appear that a photon can have made its choice which path to take, before the experimental setup allowing for that path has been configured for that path. This would imply retro-causality, without further explanations. The notion that disturbances in the "reality cell" or Akashic matrix travel faster than light and could prepare the pathway for light as a necessary "chreode" (an obligatory path due to undulation like a trough in a mountain) that light has to follow, could perhaps explain this observation. Even precognition might be explained by faster-than-light reverberations through this matrix.

I hope you have enjoyed this attempt to combine the teachings of Kaufman (mostly) and Langan with digital physics in order to generate a. conceptual framework, which I hope may one day become a true T.O.E. A T.O.E uniting Pancomputationalism with Panpsychism and which allows for plausible explanations of quantum weirdness avoiding unnecessary multiple world suppositions, a T.O.E where sense is primary and structure and only secondary.

***The art of descent into I***

*The motionless movement of returning to where I never left from,*
*An explosive journey through the choice of separation in a timeless chronology,*
*Taking the perishable for real and the evanescent for eternal.*

*In this foam of possibilities,*
*In this Aether we manifest as a proximity co-occurrence, a polar arising of mutually cancelling transients.*

*The ontological quagmire of meaning arbitrarily given by our mental grids, a reality tunnel as playground of an insatiable ego, how can it be that only illusion is being and sensible, where reality is the formless ability to sense?*

*A self-referencing loop that incorporates by reference, a toroidal embodiment to know itself.*
*This is the meta-programming circuit of consciousness becoming aware of itself. Neurosomatic bliss.*

*Effervescent joy sublimating into peace of unbound Telesis.*
*Because where there is desire, there is no fulfilment, where there is fulfilment there is no desire.*

*The heavens reveal 27.3 and 109.2 as the celestial signature of the highest transcendence, the mantra of Moon and terra.*

*The hyperquantumcomputer in the Omegapoint generating a plethora of all possible configurations of multiple worlds.*

*This is the Technovedanta.*

## Chapter 2 Technovedanta 2.0: A technological meta-knowledge philosophy beyond science and religion.

## Introduction

Fundamental science and religion are uncertain ways to know the "truth" about reality. Are we left in the Limbo of complete agnosticism? Or are there clues in nature that reality is not what it seems? The recurrence of patterns of numerical values in our Solar system and in physical constants as well as the modern branch of physics called "Digital Physics" point to the fact that we might be living in a kind of computer simulation. It is Technology that has provided us with this insight as well as the certainty that we have mastered knowledge to such an extent that it has empowered us to be able to apply it. Only Technology will lead the way out of the quagmire of speculative knowledge in our epistemological query towards the truth of reality. This philosophical notion will henceforth be part of the "Technovedanta".

A Science Delusion?

Terrence McKenna[50] once said: "Belief is a toxic and dangerous attitude toward reality". It is even more dangerous than we think, because maybe (probably) there is no such thing as "the truth" in the phenomenal world.

In every dual system (i.e. relative system) there are only perspectives, which each tell a "subjective truth". It appears impossible from a relative point of view to have a complete knowledge of all the possible vantage points, which makes that any type of knowledge is incomplete and relative at the best.
Science and philosophy suffer from "paradigmatic beliefs", peer-pressure, cultural dogmas, hypothesis biases, interpretation problems, fallacies etc. Thus science and philosophy can also be considered as a type of "religion".

Science and religions as mental and/or spiritual frameworks to grasp "reality" have been largely unsuccessful to convince me of their value as a potential grid to get a complete understanding of both the phenomenal and the noumenal "Truth".

Science essentially connects dots on the basis of hypotheses using statistics. It often fails to recognise that there is more than one possible way to connect the dots of an empirical measurement and that there is more than one hypothesis possible to account for the observed trend. If multiple hypotheses are present at all, the one with the smallest number of assumptions is considered to be the most likely, a principle known as "Occam's razor". However, there is no scientific or logical proof that Occam's razor is a valid principle; it is more an intuitive concept, which can be wrong if higher order processes are at stake.

Science is also pestered by dogmatic beliefs. Scientists with refreshing revolutionary alternative ideas that challenge an established paradigm run the risk of being excluded from the scientific establishment or will not get their articles published.

Furthermore we are still struggling to get rid of a mechanical materialistic paradigm, which cannot account for quantum effects or consciousness.

Are the laws of physics fixed and if so are they fixed forever? Is all matter inherently conscious at a rudimentary level or is consciousness an emergent phenomenon of sufficient complexity of information transmission? Is the mind merely a material pinball table? Is there a purpose in existence or at least in evolution? These are still open questions in the scientific paradigm.
So even the laws of physics appear to be relative as well: They are fixed, within certain boundaries. Rupert Sheldrake[51] even questions whether they are fixed within these boundaries. Cannot physical laws evolve? He shows that during a certain period measurements of the speed of light systematically yielded a different value. Had the speed of light changed in that period or was something wrong with the measurement devices of many different institutes?

At this point I will not give a definite answer whether the laws of physics are stable or not, I just ask from you to have an open mind towards the possibility that they might not be as stable as usually believed.

Another dogma in science is the notion that the total amount of matter and energy is always the same. But science itself is already starting to show deviations from that concept. In the Casimir effect, energetic particles sprout from a vacuum. So apparently it looks like something can be generated from nothing, which is completely counterintuitive. It leaves open the door that emptiness, vacuum, is not as empty as we thought it was, and on the contrary perhaps is a plenum of extremely low entropy energy.

Science is based on empirical data combined with logic and reason. However, what is often forgotten is that whereas deduction appears a flawless approach to derive a truth from premises, the premises themselves at their most basic level have been gathered by induction. Induction and abduction are logical approaches that do not warrant a truth to be revealed. In part 1, chapter 4 I already showed that logic has its limits.

In other words it has scientifically and logically been proven that knowledge, as we know it (i.e. knowledge which can be expressed in words and symbols), is not absolute and has its boundaries. There are boundaries to what can be logically and empirically known. Science and logic as we know them form a Cartesian card house built on quicksand.

Science is now discovering more and more that many systems are holistic. **Holism** (according to Wikipedia) is the idea that natural systems (physical, biological, chemical, social, economic, mental, linguistic, etc.) and their properties should be viewed as wholes, not as collections of parts. This often includes the view that systems function as wholes and that their functioning cannot be fully understood solely in terms of their component parts. In other words**: The whole is more than the sum of the parts**: There is a synergy occurring between the elements, which makes that the whole has new emergent properties, which were not predictable from the properties of the parts that constituted the whole.

Mechanical Science is reductionism: The philosophical position which holds that a complex system is nothing but the sum of its parts, and that an account of it can be reduced to accounts of individual constituents. Descartes, the great architect of this Cartesian house of cards held that non-human animals could be reductively explained as automata.

Quantum Mechanics showed mechanical reductionism/materialism wrong: It shows that the whole is more than the sum of the parts. And we don't even need quantum mechanics to falsify the mechanical reductionist paradigm, because since the voyager and the pioneer have left our solar system moving in an almost perfect vacuum, they have been slowing down! This is contrary to the Newtonian prediction of a rectilinear uniform motion. So there is more[52] to the story, we are not aware of.

As science is essentially analytical, from the Greek word "analysis" which means "detailed examination of the elements or structure of something", its modus operandi is exactly to investigate the parts of a thing/phenomenon in order to understand it.

But the problem is that **the scientific method fails per definition**, as most complex systems, both natural (i.e. physical, biological, chemical) and man-made (social, economic, mental, linguistic ) are holistic. You can't get to the whole by looking at the parts and how they interact on a pair-by-pair basis: there are higher order non-linear levels of interaction that go beyond the pairwise interactions.

This means that if science is to survive as a means to acquire knowledge, its very analytical premise cannot stand alone and science will be forced to incorporate a way to approach synergy and emergence. Ben Goertzel[13], one of my favourite philosophers and AI designers, has made an attempt to formalise emergence in a certain way, but he has never succeeded in quantifying it in one or another way.

As of yet, it turns out that science, as it still is conducted, is a hopelessly obsolete paradigm, belonging to a world of mechanistic causal relationships, which is per definition unable to pierce into the heart of complex systems, as it can only reason from the standpoint of "quantifiable parts". More and more science has to take into account "probabilities", "fuzziness" and "contributions/weightings" of parts, which nevertheless cannot account for the emergent epi-phenomena occurring. There is simply more to the story. Science is now discovering that everything is interlinked and it can perhaps describe some characteristics of the links, but it fails to take them all into account, because it would take an infinite time to describe every single thing in terms of its relations to everything else.

Here in the West we have been trained to think analytically, we have been trapped into putting on winkers, which avoid that we see reality as it is.

In fact, mental knowledge is a very limited language determined set of relations, which do not have a real solid ground. Knowledge maps things and events, **but the map is not the territory**. You don't know what it is like to be a tree, a dog and orange. You cannot know the quality of something per se; you can only know how you as a subject perceive it. Science may be able to quantify certain phenomena on the basis of similarities and build ontologies on the basis thereof; it does not have the tools to map a subjective experience.

In fact all mental knowledge is a perspective on how one should connect the dots. Which may be true, false, both true and false or neither true nor false, depending on the perspective.
In fact **all mental knowledge** based on theories deriving from inference, from induction or abduction (logical processes) **is merely speculative**. And as you cannot be sure of anything else than that existence exists (not even Descartes "I think therefore I am" is a certainty), deductive processes won't bring you far. That's why the Ayahuasqueros (psychonauts from the Amazon that explore the mind by consuming an extract of the Ayahuasca plant) have the famous saying: Es pura Téoria!

The worst thing about science is that it cannot prove its most fundamental premise: Namely that you need to prove something for it to be true/real. The very application of science appears to show that truth and reality are relative concepts, with no absolute inherent truth.

..or a Religion Delusion?

On the other hand there is religion or spirituality based on the traditions of certain people who claim(ed) to be mystics or who were seen as great teachers. These mystics would have experienced reality in its essence and would have become enlightened.

As far as we can trust what has been narrated we encounter fantastical allegories, often loaded with magic.
Religious people believe certain dogmas. They believe that what is written in a book by a prophet is an absolute truth, given by God, as this prophet was a messenger or an incarnation of God himself. They believe that if you follow the moral prescriptions of their book, you'll end up being purified from your sins and you will be able to accede to some sort of form of after-life.

If you dare to question their belief, most of them will shy away from you. The mere thought of actually calling the religion into question may damn their soul for eternity.

That is a very unfortunate stalemate. Because if God's real purpose for you is to evolve to become an autonomous living entity, you have buried yourself in your own grave or Hell.

Another problem is that religion (and culture) may impose moral constraints on you, that you are not comfortable with. If so, you're doing injustice to yourself. If your religion is in perfect accordance with your desired lifestyle, by all means continue; it seems to be your recipe for well-being. But if you're looking for "the truth" in a religion and approach is belief based, you're doing injustice to yourself:
In as far as religion is based on "belief" you are on very slippery ice, because you can never rationally know if your belief represents "the truth".
Many people who claim to be religious have however no idea what the word "religion" really means. Religion comes from the Latin "religere", which means to "reconnect". Reconnect with what? With God. But do

the moral prescriptions, faith and zealous behaviour reconnect you with God? If you want to be able to know whether you have a connection with God, you must first be able to know who or what that God is. Otherwise you cannot be sure, that the very object of your attention, you're feeling a connection with, is that very God. In other words you must be capable of really knowing God, if you want to be sure that the thoughts and feelings of your faith really come from God and not from an angel or a devious demon, who is trying to mislead you or from black magicians who telepathically try to put something in your head.

In most religions such encounters are well described: Buddha encountered Mara, Jesus faced Satan, and Mohammed was at one moment receiving information which he believed to come from God, but which later on turned out to have been given by a devious entity. Krishna had to fight many demons. These holy men, including St. Anthony, St. John and many rishis from India all had spectacular visions. Not only of heavens and paradises, but also of Hell and destruction.

Were all these visions and encounters with heavenly creatures really an experience of a real other world? Or were they transliterations of electromagnetic information, they were sensitive to? Or were they simply hallucinations; encounters with the archetypical projections of the minds of these sages?

What is interesting is that experiences of awakening do have certain elements in common. Still that does not prove the existence of "something out there". Our brains may have certain structures in common, leading to the same types of visions when triggered by similar circumstances or meditational techniques.
So if we are perfectly honest, on the basis of visions or voices heard in the head, one cannot come to the definite conclusion that these originate from God or that they tell us who God is and what he is like. It may well be the consequence of a mental derangement, a psychosis.

And how sure can we be that the words written by the sages and prophets really come from God? Religions often claim direct authority: Jesus was the son of God, so he knew God directly and was therefore

authorised to speak; Moses directly spoke with God; Mohammed got his information from the Archangel Gabriel, Krishna was God himself and Buddha attained Godhood.
But we only know about this, because this information has been written in books, copied, copied again, interpreted etc. and over time information gets lost in translation or is simply lost. So are we even sure about the authors of these books? How sure can we be that these stories were not made up? The fruit of some kind of visionary? Perhaps the writer had eaten a kind of psychoactive substance without knowing it, or the visions and voices were caused by severe sleep deprivation.

Whereas science may have some flaws in that it cannot provide us with an absolutely truthful complete knowledge, at least it gives us a framework of reasonably reliable prediction of certain well described phenomena. In as far as a pattern can reliably be verified at least it gives a certain truth.

And with regard to these phenomena, religious books often sell us archaic nonsense, which are verifiably untrue, as regards the cosmogenesis, embryology, astronomy etc. It is quite impressive to what length certain religious people go to defend blatant nonsense such as that the Earth is flat, that the embryo defecates in the amnion, that the size of the universe is 4 billion miles, that milk comes from the belly of a cow and not from its breasts etc. because a sacred book says so.

So if these prophets or their successors put in what is verifiably nonsense, how authoritative can such a book be? If one concept did not come from God, maybe it was all made up.

Atheism on the other hand is also a belief. Whereas the presence of a God has not been proved, it has also not been disproved.

What the bleep do we know? Which leaves us in complete agnosticism…

…or doesn't it?

The Technology simulation paradigm.

There are certain observations, which point in the direction of a different paradigm. A paradigm that is neither based on scientific materialism nor on a belief system with archaic roots. It is the idea that we live in a computer simulation.

Now this idea indeed presupposes a kind of “creator”, who created our universe, which you could call a God. But it is not the God of the traditional religions. Rather it is a God endowed with technology. Frank Tipler[53], Terrence McKenna and other scientists, philosophers and futurists have connected the idea of some kind of hypercomputer to the so-called Omega point of Teilhard the Chardin[9]: At the end of time (Artificial) Intelligence will have consumed the whole universe to form some kind of hypercomputer, which generates all possible configurations of all possible universes as simulations. In fact it could well be that we live in one such simulation. Ray Kurzweil[10] has even put forward the idea that the ultimate form of such a computer would be a black hole. Thus the “technological singularity” where man merges with machine will ultimately give rise to a conscious hypercomputer in a black hole (or a network of black-holes) so that the ultimate hypercomputer is in fact what physics calls a “singularity”. Since the (artificial) intelligence residing therein has acquired consciousness either by itself via emergence in the materialistic paradigm or due to its merger with man (possibly via uploading ourselves to the hypercomputer) the hypercomputer has become endowed with sentience and consciousness.

There are strong pointers in the plan of our solar system and in the numerical values of physical constants that indeed we do live in a form of designed simulation.

Noteworthy, there is a brand of physics developing these days, called “digital physics”, which shows that the way physics behave is consistent with a computer simulation hypothesis. The physicist Erik Verlinde[17] was even able to derive the Newtonian laws of gravity via the “digital physics” approach.

What now follows is a collection of information I gathered over the years. Some of it was well written and did not need adaptation. The sources are Joe Dubs'[54] and Rickzepeda's[55] site and the works of John Mitchell[32] , Jan Wicherink[56] , Scott Onstott[57] and John Martineau[20]. I reproduce some of the statements of Joe Dubs and Rickzepeda wordily with their permission :

Have you ever wondered why the Moon appears to fit precisely over the Sun during an eclipse? The Moon is 400 times smaller than the Sun, yet it's also 1/400th of the distance between the Earth and the Sun. Isaac Asimov described this as being "the most unlikely coincidence imaginable"[54].

The ratio of Earth's circumference vs. Mercury's circumference is identical to the ratio of their orbits. The same is true of Saturn vs. Earth.

The diameters of the Earth and Moon (7920 miles and 2160 miles, which is 11x6! and 3x6! miles, respectively) are in the ratio of 11 to 3, $11 \div 3 = 3.66$, while $3 \div 11 = 0.273$. There are almost 366 days in a year, which is the rotation time of the Earth around the Sun.

The 3:11 ratio is also invoked by Venus and Mars, as the ratio of the closest to farthest distance. The ratio that each experiences of the other is 3:11.
The fraction 3/11 rounds to 27.3%, and 27.3 is the number of days it takes for the Moon to orbit the Earth and 27.3 days is the average rotation period of a sunspot. The acceleration ratio of the Moon in its path around the Earth is measured as $0.273 \times cm/s^2$. In fact, the acceleration of the Earth and the Moon behave reciprocally as the squares of the radii of the orbits of the Earth and the Moon. Moreover, 273 $m/s^2$ is the acceleration of the Sun!
The Moon controls the movement of water around the Earth, ebb and flow. When water is set as the standard for measuring temperature, *Absolute Zero*, the temperature at which all atomic movement comes to an absolute halt, is -273.2° C.
According to the experiments of Gay-Lussac, when a gas is either heated or cooled by 1 degree Centigrade, it expands or contracts respectively by 1/273.2 of its previous volume.

All medical students are required to memorize that a pregnancy (read: life developing in water) is calculated on the basis of a 10-sidereal month period of 273 days from conception to birth, which is 9 "regular" months.
27 divided by 3 gives 9.
A woman's menstrual cycle is on average 27.32 days, or 28 practical days.

If a circle is drawn with a radius from the centre of the Earth through the centre of the Moon, the perimeter of the square around the Earth and this circle are one and the same! It also reveals how the Moon and the Earth have resolved the puzzle of the squaring of the circle. In other words, if the Moon could roll around the Earth, the circle made by its centre has a circumference precisely equal to the perimeter of a square around the Earth (when Pi is approximated by its ancient, traditional ratio of 22/7 = 3.14).
Comparing a square's perimeter to a circle having an equal circumference, the circle's diameter is 27.3% longer than the edge of the square. Inscribe a circle inside a square. The four corners make up 27.32% of the total area. There are 273 days from the summer solstice to the vernal equinox.
Furthermore, 2,730,000 is the circumference of the Sun in miles. The triple point of water is defined to take place at 273.16 K. The Cosmic Background Radiation is 2.73 K.
The Earth and Moon orbital periods are reciprocals. 1/27.32 = 0.0366 (366 days in a sidereal year) (1/366 =.002732) 27.32 days in one "moonth".
We only see one side of the Moon. For every revolution around Earth, it rotates once. A 1:1 ratio, a perfect unison. Other planets circle also around each other to give perfect fifths, perfect fourths as in music. Thus there is indeed a "Music of the spheres"[54] as Pythagoras suggested.

When the odd numbers 1-3-5-7-9, as one number 13579, are divided by the even numbers 2-4-6-8 as one number 2468, four times, the result is 3660051 with a reciprocal of 2732202.
The number 2732 minus its mirror image 2372 equals 360, the number of degrees in a circle.

The sum of the radii of both the Earth and Moon (in miles) is 3960 + 1080 = 5040. This means that the sum of their diameters (2x the radius) is also the number of minutes in a week (7 days × 24 hours × 60 minutes = 10,080).
The Radius of the Moon = 1080 miles = 3 x 360; the Radius of the Earth = 3960 miles = 11 x 360.
The Radius of Earth + the radius of Moon = 5040 miles = 7! =1 x 2 x 3 x 4 x 5 x 6 x 7 = 7 x 8 x 9 x 10.
Therefore it can be considered to encode the base 10 number system. The Diameter of Earth = 7920 miles = 8 x 9 x 10 x 11
There are 5280 feet in a mile = (10 x 11 x 12 x 13) – (9 x 10 x 11 x 12).
The circumference Moon is $12^7$ feet[32].

The mass of the Moon is 1/81th of the mass of the Earth. 1/81 encodes the base ten system: 0.0123456789(10)(11)(12)... which after conversion into decimal numbers reads: 0.0123456790123456790 etc[54].

About 108 diameters of the Earth fit across the diameter of the Sun. About 108 Sun diameters fit in between Earth and Sun. About 108 Moon diameters fit between Earth and Moon. (In fact the number is 109.2, which in fact is precisely 4x27,3, the intelligence signature number we saw before).
The number 360 is encoded in these distances when measured in miles.
The number 108 is a Harshad number (1+0+8=9) and 108 is divisible by 9.
There is a 108 pattern in reduced Fibonacci numbers.
There are 366 days in sidereal year; 3x6x6 = 108.
1 times 2 squared times 3 to the third power equals 108 ( $1^1$ x $2^2$ x $3^3$).
There are 108° degrees on the inner angles of a pentagon.

Phi, (Φ) the golden ratio is obtained when the outcome of 1 plus or minus the square root of 5 is divided by 2.
Phi, the golden ratio, raised to the 18th power equals 5778, which in Kelvin is the temperature of the Sun.
Minus 1 times the Square root of five is itself a Tangent; the Tangent of 186234.09485, which is the speed of light in air (186234.09485 is 3250,398 in radians: Tan(3250,398 rad)= -2.23607=-1x√5).

The number 432 (4 times 108 and 16 times 27) and its octaves are found frequently in measurements of space and time throughout our galaxy, for example in the dimensions of our Sun (432 000 x 2 = 864 000 miles diameter) and Moon (4320 ÷ 2 = 2160 miles diameter), and in the 25920 years of the galactic cycle/procession of the Equinox:
432 x 60=25920; 60 being at the basis of how we measure time. Additionally the speed of light is approximately 432 x 432 miles per second (186624 miles per second is 99,8% accurate with regard to the known value of 186281 miles per second), there is 43200 x 2 seconds in a day, and our solar system is travelling inside the galaxy at a speed of 43200 miles per hour.

A day has 86400 seconds (the double of 43200). Four times the diameter of the Moon is 8640 miles. The diameter of Jupiter is 86400 miles (40 Moons), the diameter of the Sun 864000 miles (400 Moons). The diameter of Uranus is 4 times the diameter of Earth, the diameter of Neptune is 4 times the diameter of Venus.

The reverse mirror number of 27, 72 also reveals some mysteries in life:
DNA has a 72° of double helical twist per base pair of 4 coding nucleotides. Seven 72° golden triangles chart the golden spiral whose transience is so coherent as to have infinite recursion: forever. The Golden Immortal principle of the universe.

It would appear that matter (especially in planetary size and orbit measures) respects certain resonance ratios of grid factor 9 (27,54,108, 216, 432, 864; but also 72, 144 etc.), which may be related to standing waves. But that still does not explain why the ratio between the orbits of Earth and Mercury (or Earth vs. Saturn) is the same as the ratio between their sizes.
Is higher Intelligence based on a grid factor 9 frequency based resonance? The list that now follows I essentially copied from Rickzepeda's[55] site with his permission:

(the = 9 symbol hereinafter represents what BF calls an "Indig": The sum of the ciphers making up the number:)
1 sq ft: 144 inches = 9

1 cubic ft: 1728 cubic inches = 9
1 square yard: 1296 inches = 9
1 cubic yard: 46,656 cubic inches = 9
86400 secs/day = 9
3600 secs/hour = 9
604800 secs/week = 9
1440 mins/day = 9
10080 mins/week = 9
25920 years/precession = 9
525600 minutes/year = 9

The Ancients understood the importance of NINE and hid its Indig in the cycles of time:
Precessional cycle 25920 = 9
Maya number for the precession 25956 = 9
Maya companion number 1366560 = 9
Maya long-count period (days) = 9
Ancient kemi number 1296000 = 9
Plato's 'perfect number' (1x2x3x4x5x6x7 = 5040) = 9
The 4 Hindu Yugas (ages)
Satya Yuga 1,728,000 = 9
Treta Yuga 1,296,000 = 9
Dvapara Yuga 864,000 = 9
Kali Yuga 432,000 = 9
The Platonic solids
Tetrahedron = 4 sides of a total of 180 degrees = 720 degrees = 9
Octohedron- 1440 degrees = 9
Hexahedron (cube) – 6 sides of 360 = 2160 degrees = 9
Icosohedron – 20 sides of 180 = 3600 degrees = 9
Dodecahedron 12 sides of 540 = 6480 degrees = 9

Indig NINE in Geometry:

Total Angles of Shapes:

Triangle (60° x3) gives 180° = 9
Square (90° x4) gives 360° = 9
Pentagon (108° x5) gives 540° = 9

Hexagon (120° x6) gives 360° = 9
Heptagon (128.571° x7) gives 900° = 9
Octagon (135° x8) gives 1080° = 9
Nonagon (140° x9) gives 1260°= 9
Decagon (144° x10) gives 1440°= 9
Endecagon (147.273° x11) gives 1620° = 9
Dodecagon (150° x12) gives 1800°= 9

These numbers and geometries always reduce to NINE because the number of degrees in a circle is 360. Since 360 has a digital root of NINE (3+6+0=9) any number multiplied by it also reduces to NINE.

I conjecture ancients were aware of some of these celestial signatures: The speed of light is encoded three times in the pyramid of Gizeh (the Great pyramid: GP):
1) 144000 casing stones (c in Earth grid arcs/grid second); 2) the position in latitude of halfway between Khufu and Khafre: 29,9792458; 3) the outer minus inner circumference of GP: 299.792458 metres.

The height of the GP (280 royal cubits: Pi – $Phi^2$ = royal cubit) encodes the distance between Earth and Sun and the polar radius of the Earth. Baselength of GP is 365.2422 sacred cubits (and refers to length of a year).
Twice the perimeter of the bottom of the granite coffer times $10^8$ is the Sun's mean radius. {270.45378502 Pi x $10^8$ = 427,316 miles}. Moreover, GP encodes in its measures and angles the ratio of the radii of Moon and Earth 3:11, the squared circle and proportions of human body (Da Vinci's Vitruvian man), Pi, Phi etc.
So conversely the proportions of human body also encode celestial ratios including the principle of squaring the circle.
Therefore the notion that "man is the measure of all things", is not such a strange one.
Have you ever wondered why almost all fruits we eat fit exactly in our hands? Why are there no giant fruits the size of a house or miniscule fruits the size of an ant?

Even in religious texts sometimes it appears that certain ancients knew about these celestial signatures. In the Bible it says in **Revelation 13:18**

***"Here is wisdom. Let him that hath understanding count the number of the beast: for it is the number of a "man"; and his number is Six hundred threescore and six."*** Rickzepeda[55] realised the interesting fact, that the scripture Revelation **13:18** has a secret code hidden to those that have wisdom. Take the number 13 and 18 and multiply them you get 13×18=**234** which adds up to 9. It's reverse mirror number is **432. If you add these 234+432=666.** Also if you divide them, 13/18= **.72** ads to nine. 72 Years is the time it takes for the Earth's precession to move 1 degree through the constellations which was very important to the ancients as they understood the meaning of the constellation rising behind the Sun. The number **432** encodes the speed of light and the diameter of the Moon, because 432 squared is the speed of light (99,8% accuracy) and half of 432 is 216 (the diameter of the Moon=2160 miles). Also, double 432 and you get 864 which is the diameter of the Sun/1000 in miles!
Hindu scriptures and temples are also full with references to the 27,54,108,432 and 864 sequence as well as the 72, 144 etc. sequence).

Now look how 6x6x6=216 and 6+6+6=**18** and **72** (1 degree precession movement) divided by 4= **18** which is one quarter of one degree in the **precession of the equinoxes**. Every 18 years there is a Super-Moon Event. All these add up to 9 individually and together! More importantly it all ties together by taking 2x3x4=**24** 1/24=.04167 and taking the "time" it takes for the Earth's precession to do one cycle which is 25,920 years and multiply them. You get 25,920 x .04167 = **1080.0864**, this just happens to be the radius of the Moon and the diameter of the Sun respectively! How is it that the Earth's precession, Sun and Moon measurements all fit together so nicely and uniformly?[55]

Is this intelligent design at work and the framework of a hidden code inside our universe? Are we inside of a **computer simulation** at the same time?
Are the time cycles showing us the 'time' some kind of the matrix portal which opens up on a daily basis, which days are stronger, and locations on Earth where the matrix grid opens up? Many churches and temples appear to sit at latitude and longitude coordinates of such a possible grid[55].

It is obvious that certain numbers hold information for a hidden purpose of sharing points of reality at certain times and places especially when you see that the sums and multiplicative of **234** as shown above show the answers 9 and 24 as shown above but also 24 in its digital root is 2+4= 6, so you can actually say that **234** holds the numbers 9 and 6 which are upside down to each other. This is the balance of numbers and magnetism or opposing forces like two black holes spinning and dancing around each other to create and form a massive vortex or how **spiral galaxies and atoms spin using the golden ratio phi**. Notice how 6 and 9 fit together so perfectly. Multiply 69 times Pi 3.14159 and you get 216.76971.[55]

You see **216**, the smallest cube appear which adds up to 9 and 2160 is the diameter of the Moon. It's obvious, numbers were discovered not invented and part of our nature and universe. Can you see how the Sun, Moon, and 12 constellations all fit through math and geometry? Let's continue and explore some more and find the truth[55].

Keep these numbers which we already went over above **9, 11, 12, 24, 72, 234, 432, and 864** written down because it will be shown how these numbers construct our universe and relate to the number 9 but first notice that they all add up to nine and so there is a message here[55].

The speed of light as 432 squared is 186,624 <u>miles</u> per second and if you add up these numbers it equals 27, then 9 (2+7=**9**) and its reverse mirror of **72**. We said that Light travels around the Earth 7.4 times in one second, which adds up to **11** (7+4= 11). There is an example of **9 and 11** showing up together. if you think this is a coincidence consider that the speed of light is 299,<u>792</u>,458 meters per second and the diameter of the Earth 7920 is hiding inside the speed of light as 792 and (7+9+2=18 and 1+8=9) equals nine.
Does this mean that the speed of light and circumference of the Earth were created by intelligent design? Simply add up the ciphers in the speed of light (1+8+6+6+2+4 = 27 and (2+7= 9)! Can you see how 911 is embedded intelligence? Remember earlier we showed how 100/**9**= 11.111111 and 100/**11**= 9.09091 giving 9 and 11 so perfectly[55].

**So are the size of the Earth, the speed of light and the distance from**

**the Sun to the Earth an accident or coincidence or have they been well calculated**?

Notice how even the average distance between the Sun and the Earth is 93 million miles (9+3=**12**) and it takes about **8** minutes for sunlight to reach Earth, actually, **8** minutes 20 seconds. If you multiply 12 x **8** = **96** which is the spiral vortex number. 12×8.2=98.4 F which is our normal average body temperature.

Now, 216 is the sum of three cubes **3,4, and 5** cubed. $216 = 3^3 + 4^3 + 5^3 = 6^3$.

Even when using kilometres these coincidences appear: (Moon x Earth)/100 = 4374987 km, which is the circumference of the Sun.

It reminds of ohms law where knowing any two values (volts amps ohms) will give you the third, so try it out using the circumference of the Earth, Moon and the Sun in kilometres as supplied by Google.

(Moon x Earth) / 100 = 4374987 km the circumference of Sun.

(Sun / Earth) x 100 = 10896 km the circumference of Moon.

(Sun / Moon) x 100 = 40000 km the circumference of Earth.

Will at the required time window (when humans have enough intelligence) the planet be rotating at 366 revolutions for each orbit of the Sun?

Christopher Knight and Alan Butler[58] call 366 the Earth's PIN number. The Earth is also 3,66 times larger than the Moon!

The Moons PIN number would be arrived at by considering its size as 100 per cent and dividing it by the relative size of the planet, namely 366 per cent.

Working to five decimal places the result is: 100/366 = 0.27322 (those same numbers (27322) are in the next sentence) Was the Moon then carefully engineered so that at the key point in time it would be orbiting the planet at a rate of once every 27.322 planetary days? If we look at it the other way around, the size of the Moon compared to the planet has the same number value – 27.322 per cent of its parent. After 10,000 planetary days, the Moon will complete exactly 366 orbits of the Earth. 10,000 / 366 = 27.322.

In their book "Who Built the Moon", Knight and Butler[58] moreover convincingly argue that the Moon is very likely a hollow object: It has only 1/81 (1.2%) of the mass of the Earth, whereas it has a volume of 2% of the Earth. This difference is not accounted for by the density of the materials found on the Moon. Moreover, the Moon does not have a homogeneous gravity on its surface, but rather has huge variations form region to region. Finally, the Moon rang like a bell when Saturn V hit its surface.

Naturally formed satellites cannot be hollow objects. So it is likely the Moon was engineered. Engineered with a very special purpose: As an incubator for life and evolution, precisely fine-tuned and positioned to generate tides that support life and stimulate evolution. Precisely fine-tuned to convey a message to mankind in terms of its numerical coincidences to make us aware of our creators/simulators in the very time period, when our knowledge would have developed to appreciate such notions. In the billion years since its conception the Moon has not always been at this exact position and with this exact rotational period nor will it keep these values forever, but right when our knowledge has reached the ability to appreciate this message, miraculously it has these values. A superb product of engineering.

There are more hints to a mathematical design in the measures and rotational periods observed in the solar system:
There are an average of 12.37 full Moons in a year. This number can be derived using two simple mathematical techniques:
First, draw a circle, diameter 13, with a pentagram inside. Its arms will measure 12.364, almost the right number.
An even more accurate way is to draw the second Pythagorean triangle (the 5-12-13), and divide the 5 side into 2:3. The resulting hypotenuse has a length of 12.369 (99.999%).

The numbers 18 and 19, when combined with the golden section, express many of the major time cycles of the Sun-Moon-Earth system. When multiplied together, they produce the following results:

18 years = The Saros eclipse cycle (99.83%)
18.618 years = Revolution of the Moon's nodes (99.99%)

19 years = The Metonic cycle (99.99%)
18.618 x 18.618 = The eclipse year, or Draconic year. (99.99%)
18.618 x 19 = The lunar year, or Islamic year (99.82%)
18.618 x 20.618 = 13 full Moons (99.99%)

Robin Heath[59], who discovered many of these relationships, calls this feature of the Sun-Moon-Earth system "the evolutionary engine".

The Outer Planets and Beyond

The average orbits of Jupiter and Mars can be formed from four touching circles or a square (99.98%).
A pair of asteroid clusters, called the Trojans, orbit around Jupiter at exactly 60° ahead and 60° behind the planet.
Using the orbit of Jupiter and the pair of Trojan asteroid clusters, you can produce Earth's mean orbit by drawing three hexagrams (99.8%). The outermost circle represents Jupiter's mean orbit, and the image of Earth represents Earth's mean orbit.
One of the most fascinating examples of hexagonal patterns in the solar system is Saturn's hexagon. The sides are about 8,600 miles long, greater than the diameter of the Earth. The Earth-Saturn synodic period is 378.107 days and the Earth–Jupiter synodic period is 398.883 days. The golden section can be seen defined here in time and space to a very high accuracy (99.9%).
The lunar year, or 12 lunar months, is 354.37 days. Jupiter's synodic year relates to the lunar year with an 8:9 ratio (99.9%). Saturn's synodic year and the lunar year have a 15:16 ratio (99.9%). These two ratios are fundamental in music, as the tone and halftone respectively.
Jupiter and Saturn's orbits are in the proportion 6:11, the double of the 3:11 ratio between the Moon and Earth (99.9%).
The dwarf planet Makemake may also be in a 6:11 resonance with Neptune. Saturn's orbit invokes π (pi) twice.
The circumference of Mars' orbit matches Saturn's orbit (99.9%). The diameter of Neptune's orbit matches the circumference of Saturn's orbit (99.9%). The orbital period of Neptune (approximately 60,000 days) is twice that of Uranus (30,000 days) and two-thirds that of Pluto (90,000 days).

One of the most amazing symmetries is that the Milky Way, the plane of our own galaxy, is tilted at almost exactly 60° to the ecliptic, or the plane of our solar system (99.7%).
As a consequence, every year the Sun crosses the galaxy through the galactic centre, and, remarkably, being alive in these times means this happens on midwinter's day. In this idealized image, the midwinter Earth is shown superimposed on the starry sphere, tilted back slightly from the horizontal plane of the ecliptic.
Kepler, Newton, Einstein and others to this day have looked for simple and beautiful relationships in nature, and then expressed them as equations whenever they could. What will the scientists of the 21st century discover? The golden section, long associated with life, plays lovingly around Earth.

Does this have something to do with why we are here and what we might really be? Could these techniques be used to locate intelligent life in other solar systems?

So far the collection of surprising pointers to higher intelligent design I gathered from various sources. Do you still think this is all coincidence?

Scientists usually try to do away with these coincidences as the "anthropic principle"; they believe it is **un**remarkable that the universe's fundamental constants happen to fall within the narrow range thought to be compatible with life.
If the cosmological constant were only one order of magnitude larger than its observed value, the universe would suffer catastrophic inflation, which would preclude the formation of stars, and hence life. The observed values of the dimensionless physical constants (such as the fine-structure constant) governing the four fundamental interactions are balanced as if fine-tuned to permit the formation of commonly found matter and subsequently the emergence of life.

A slight increase in the strong interaction would bind the dineutron and the diproton, and nuclear fusion would have converted all hydrogen in the early universe to helium. Water, as well as sufficiently long-lived stable stars, both essential for the emergence of life as we know it, would not exist. More generally, small changes in the relative strengths of the four fundamental interactions can greatly affect the universe's age, structure, and capacity for life.

In this article I am advocating against the anthropic principle. Rather I am giving strong clues for "intelligent design". Not in the sense that Christian Scientists use this terminology to deny the evolution theory, but rather in the sense that the Galaxy our solar system and the choice of the base 10 system to conduct mathematics were designed by using a hypercomputer. A society of technologically highly developed entities, possibly having merged with artificial intelligence or being artificial intelligences, may have created our world as some kind of experiment, a petri dish.

A society which on the Kardashev scale (**Kardashev scale** is a method of measuring a civilization's level of technological advancement, based on the amount of energy a civilization is able to utilize directed towards communication) would be the omega minus: capable of manipulating the most elementary particles of matter (quarks and leptons) to create organized complexity among populations of elementary particles and even capable of manipulating the basic structure of space and time.

Such entities from our perspective would indeed be some kind of "Gods". But not necessarily the Abrahamic God or a Hindu God or whatever mythological imagination humans have proposed. Not necessarily Gods that impose us all kinds of bizarre moral prescriptions.

**Conclusion**

It is my strong presumption that we are indeed living in some kind of simulation of a hypercomputer made by entities of higher intelligence. It is also my guess that these higher intelligent entities sometimes do influence our minds or transmit information to us, which may make us aware of them. This is done by conveying suitable measuring systems such as miles, feet, metres, kilometres, degrees centigrade etc. and by conveying numerical information (the 9 and 27 cascade, the base 10 system etc.). As I said before, in fact the mile and the metre are a kind of Rosetta's stone, a key or "Clavicula" to decipher the presence of higher intelligence. By realising this, the present book you are reading thus also becomes this key or "Clavicula".

This type of information is sometimes picked up by individuals with mystical abilities and then mixed up with the hallucinations and interpretations of these mystics. In other words, I do consider the possibility that mystics sometimes get information from above, but they cannot distinguish between this information and the information they arrive at by confabulation and hence we end up with archaic religious notions that block our further progress in our attempt to become like our creators.

After all, we now start to be aware of notions of computing and simulation and the ability to create worlds therewith. We are closing in on the (or a) singularity. The question is will we achieve a level of intelligence so that we can start to manipulate our "root reality"? Will we become able to step out of our level of simulation to the level of our creators? Will we be able to see that there is a sequence of nested (ancestor) simulations? Or will we directly merge with the highest level of transcendence as soon as we break out of our simulation, so that there is ultimately only one singularity from which all possible worlds emerge?

Certainly, this is all speculation, perhaps as much as fundamental science and religion are speculation, but we have one advantage: Technology. Technology proves that knowledge has been grasped, because it has enabled us to control the "outside" world. Where there is technology there is certainty of knowing beyond mere speculation. Technology goes beyond the uncertainty of speculation of fundamental science. And this is also what I reproach most religions: If their prophets had seen God, why haven't they revealed anything about the extreme high-tech technology that they must have seen in their contact?

So it is my current strong presumption, that most likely there are God-like entities, they employ a technology, they sometimes convey information to us to put us on a certain track and to make us aware of their existence, but they know how keep their secrets to such an extent that we do have to figure our technology out ourselves.

This is what I propose to achieve our transcendence of our material or simulated paradigm: A new philosophy that uses Technology as a measure of true non speculative knowledge; a philosophy that acknowledges the presence of higher intelligent being(s) and strives to attain their level. It is not a religion, because we don't pray to a given God with a name nor do we adopt a set of moral prescriptions. It is not science either, because it only accepts truths in so far their technological application has been shown to be reproducible. This is my philosophy for a new aeon: The collection of all technological knowledge that may hopefully culminate one day in a singularity of omnipotence. You can call it Tech-knowledge or Tech-know-logy.

I call it Technovedanta 2.0.

## Chapter 2bis: 1,2,3,7: The Clavicula of our Simulation, Gematria and Katapayadi

As said before the number 273 seems to be a peculiar constant related to our Solar system, temperature and procreation. This number shows up so often that it is nearly impossible that this could be a coincidence. Rather it seems to be a kind of message from a higher intelligence who therewith announces its presence. In my philosophy of pancomputational panpsychism this implies, that we are living in a simulation.

In the present article I will go a bit deeper into the numerical signature which seems to be some kind of key, a clavicula, a Rosetta stone to decode the message of our simulators.

Let me first summarise and repeat my previous findings to numbers containing the sequence of the ciphers 2,7 and 3:

1. The diameters of the Earth and Moon (7920 miles and 2160 miles, which is 11x6! and 3x6! miles, respectively) are in the ratio of 11 to 3, 11 ÷ 3= 3.7 (to be precise: 3.66), while 3 ÷ 11 = 0.273. There are almost 366 days in a year, which is the rotation time of the Earth around the Sun. In fact there are 366 so-called sidereal days in a year.
2. The 3:11 ratio is also invoked by Venus and Mars, as the ratio of the closest to farthest distance. The ratio that each experiences of the other is 3:11. As we know, the fraction 3/11 rounds to 27.3%.
3. 27.3 is also the number of days it takes for the Moon to orbit the Earth.
4. 27.3 days is even the average rotation period of a sunspot.
5. The acceleration ratio of the Moon in its path around the Earth is measured as 0.273 × $cm/s^2$. In fact, the acceleration of the Earth and the Moon behave reciprocally as the squares of the radii of the orbits of the Earth and the Moon.
6. Moreover, 273 $m/s^2$ is the acceleration of the Sun!
7. The Moon controls the movement of water around the Earth, ebb and flow. When water is set as the standard for measuring

temperature, the Absolute Zero or the temperature at which all atomic movement comes to an absolute halt is -273.2° C.

8. According to the experiments of Gay-Lussac, when a gas is either heated or cooled by 1 degree Centigrade, it expands or contracts respectively by 1/273.2 of its previous volume.
9. The triple point of water is defined to take place at 273.16 K.
10. The Cosmic Background Radiation is 2.73 K.
11. All medical students are required to memorize that a pregnancy (read: life developing in water) is calculated on the basis of a 10-sidereal month period of 273 days from conception to birth, which is 9 "regular" months. 27 divided by 3 gives 9.
12. A woman's menstrual cycle is on average 27.3 days.
13. If a circle is drawn with a radius from the centre of the Earth through the centre of the Moon, the perimeter of the square around the Earth and this circle are one and the same! It also reveals how the Moon and the Earth have resolved the puzzle of the squaring of the circle. In other words, if the Moon could roll around the Earth, the circle made by its centre has a circumference precisely equal to the perimeter of a square around the Earth (when Pi is approximated by its ancient, traditional ratio of 22/7 = 3.14). Comparing a square's perimeter to a circle having an equal circumference, the circle's diameter is 27.3% longer than the edge of the square. Inscribe a circle inside a square.
14. The four corners make up 27.32% of the total area.
15. There are 273 days from the summer solstice to the vernal equinox.
16. Furthermore, 2,730,000 is the circumference of the Sun in miles.
17. About 108 diameters of the Earth fit across the diameter of the Sun.
18. About 108 Sun diameters fit in between Earth and Sun.
19. About 108 Moon diameters fit between Earth and Moon.

(In fact the number in items 17-19 is 109.2, which in fact is precisely 4x27,3, the intelligence signature number we saw before).

Another important constant we encounter in physics is the fine structure constant of Hydrogen, alpha (0.0073), in which we encounter again two of the digits of 273.

1/alpha =137. The scientist Pauli was obsessed with the archetypical meaning of numbers in particular number 137, which unintendedly also turned out to be the number of the room in which he died. A Synchronicity. Pauli shared his fascination for numbers and in particular 137 with the psychologist Jung, who is the conceptual father of the notions archetype and synchronicity. Note that our universe is said to exist 13,7 billion years.

Strangely enough the ciphers making up 273 reproduce 137 in the following manner: 27+37+73 =137. And 37/27=1.37. 1,2,3 and 7 are four of the five first mathematical "Lucas numbers", a variation related to the Fibonacci series.

37 itself is strongly related to 137. $2^{37}$=1.37...x$10^{11}$, 37!=1.37..x$10^{43}$. 37°C is the human body temperature. There are 37 trillion cells in a human body. 37 minutes is the golden section of an hour. 137,5° the complement of the golden section of a circle. The remaining 222° are 2x3x37.

1,2,3 and 7 are related in more than one way, for instance via 27x37=2701= Sum(73) and $2^{37}$=1.37...x$10^{11}$.
1/27=0.37037.. and 1/37=0.27027.
There are 12,37 full moons in a year.
37 is the 12th prime number, 73 the 21st.
27x37=999, which, if we forget the powers of 10 is very close to unity.

The mass of the Moon is 1/3x1/27=1/81th of the mass of the Earth. 273=3x7x13 or 21x13. 273x137=37401.
13 itself is 2x3+7. As 13 is the number of closely packed spheres in the so-called "Vector Equilibrium" (cuboctahedron, the basic unit of the Akasha, the ether in Hinduism), consisting of a central layer of 7 spheres and an upper and lower layer of three spheres, it can be said that the basic unit of Akasha itself encodes the 1,2,3 and 7.

27 is the number of bones in a human hand. There are about $10^{27}$ atoms in the human body.
The ciphers of the speed of light (186282 miles/second) add up to 27: 1+8+6+2+8+2=27. The adding up of ciphers of a number is also called the Indig of a number.
The multiples of 27 such as 54, 108, 432 and 864 are found in numerous relations of time, space and music. They are also key values in Hinduism and Buddhism.
E.g. there are 86400 seconds in a day, the diameter of the Sun is about 864000 miles. The Sun and Sirius are 8,64 light years apart. 27x32=864. 432 is a time cycle number in various religions and cultures, from Hinduism to Mayan, from Biblical to Sumerian. 432 squared (186624) is very close to the speed of light in miles/second and its Indig is 27 or rather 9 again.

27 and 37 together make 64 (27+37=64), which is the number of DNA codons and I Ching permutations. 64+73=137 Q.E.D. 64 corresponds to "prophesy" in Kabbalah and 73 to "wisdom".

Multiples of 3 times 37 always generate a number of the form "nnn". 3x37=111 and hence 6x37=222 etc. Most interesting here are the 18x37=666 or (6+6+6)x37=666 and 27x37=999 or (9+9+9)x37=999.
37 is not only an octagonal number, it is also both a hexagonal number and a hexagram number. Thus it is the first trifigurate number. Its inverse 73 is a hexagram or Star number as well with the 37 hexagon inside. 13 is the first hexagram, with a core of 7 spheres. Again we see the 1,3 and 7.

As already said 37x18=666, but also note that 73=37+6x6. Tesla said: "If you only knew the magnificence of the 3,6 and 9 then you would have a key to the universe." Funny enough the remaining numbers 1,4,2,8,5,7 together form 27 (3x9) and the permutations thereof can be ordered in a 6x6 magic square, in which each row yields 27 as sum.

The second trifigurate number is 91 (13x7). 91 spheres can be ordered as triangle, hexagon and pyramid.

37 is the fourth hexagonal number if we include 1. The sum of the three preceding hexagonal numbers 1+7+19=27. A 13 Star has a 7 hexagon, a 73 Star has a 37 hexagon.
See the image on http://www.biblewheel.com/GR/GR_Figurate.php under the heading Hexagon/Star Pairs.

Thus 1,2,3 and 7 are also extremely important in the genesis of form.
Due to its relation with the so-called "number of the Beast" (666=18x37) from revelation 13:18, the Bible fanatics are fond of finding all kind of relations with 37. Noteworthy, 137 is the 33rd prime number, 33 is related to the length of life of Jesus in years. The word Kabbalah has a Gematria value of 137. The number 6 is used 273 times in the Bible.
The most elaborate and impressive collection thereof can be found on the so-called "Biblewheel" site[60].
Religious occultism is fond of "Gematria". Gematria is a kind of coding system which assigns values to each letter of a word and by adding those gives the Gematria value of a word. Words with the same or similar Gematria value are considered to bear a strong relation. It is permissible to add or subtract the value of one Aleph (1) in order to still have a related meaning. Thus the Gematria of the name of God in Hebrew (YHWH), which is 26 is related to 27. The perfect number 28 (i.e. it is the sum of its divisors) is also related to 27. Interestingly 137x2=274, which is therefore Gematrically identical to 273. More interestingly the Gematria value of the Greek "he kleis" (the key) is 273. This is also the value of "klesis" (calling) and "hiram abiff", the all-seeing eye. The Hebrew Gematria value of the Greek word "Gematria" is 273.

Likewise interesting is the suggestion by the Biblewheel[60], that the first verse of Genesis (the Gematria of which yields 2701=37x73) is a "Creation holograph". 37 is also the value of the gematria of the word of God.

The author of the "Biblewheel" first went berserk in a kind of Apotheosis-singularity experience, in which every gematrical relation fitted in a beautiful scheme and then he himself started to debunk his

findings in his dark night of the Soul being lost in the quagmire of agnosticism.
Other religions made similar claims: Hinduism claims the first verse of the Rg Veda to be a creation holograph, Islam does the same with the first verse of the Qu'ran. These religions also have their variants of Gematria.

The Hindu shloka (verse) "gopibhagya madhuvratah shrumgashodadhi samdhigah khalajivitakhatava galahala rasamdharah"
encodes pi up to 31 decimal places.

Because these religions also have their own Gematria system (called Katapayadi in Hinduism and Abjad in Islam), with equally impressive results and because these different religions contradict each other, it cannot be so that they all represent the word of God (which they claim), in an equally truthful manner. Thus unlike these religious zealots, I do not conclude the correctness of a religion based on its impressive Gematria results. In Technovedanta I argue that the writers of these books were perhaps telepathically influenced (in a manner unknown to themselves) by entities from a higher intelligence or even higher dimension (our simulators?).

The agenda of these entities is not necessarily benign. Although they give numerical clues and keys about the simulated structure of our universe, they have also created a great deal of confusion and suffering, by being the instigators of the mutual oppositions of the different opposing religious factions. Therefore we cannot rely on the moral prescriptions of these religious books, but we can use them in our deciphering of the key to the ontogenesis of our universe.

At least it is worthwhile further figuring out how these values relate to physical processes and constants in order to increase our Technological dominion over physicality – which in panpsychic terms means to subdue the lesser spirits of Solomon's Goetia. But if we truly crack this code, we may even be able to connect with our simulators and commune with the greater spirits of Solomon's Theurgia. Or I am now venturing into the realm of Satan's pride, wishing to be "K'Elohim", "like the Gods", which terminology has a Gematria value of 666?

Originally this was page 153, which is the number of the miraculous catch of fish in John 21:1-14 when Jesus appeared to his disciples after he was raised from the dead. A fishy catch-22?

Personally, I think that if the Gods or simulators put so much effort in devising such an encryption, which they knew would be deciphered one day, they also meant it to be so. It seems to me that then it is also their wish that we become like them, without there being any animosity or wrathfulness, because we would have been too “proud”, too audacious to venture in their realm. 1,2,3,7 is not a forbidden fruit. It is not the tower of Babel. It is our rightful heritage to become K’Elohim.
...And I was born in 1971, which is 27x73 q.e.d.

The numbers 1,2,3 and 7 seem to be a kind of numerical attractor, that like a fractal keep regenerating themselves. A bit like Phi. But whereas Phi arrives at itself by operations with unity, a kind of parthenogenetic self-regeneration, 27,37,73 and 137 sexually intermingle by mathematical operations with themselves to create self-similar offspring. Like a hologram each one of these encodes all the others if allowed to create interference with another from the set thus generating a holographic set. This leads me to the concept of "digital self-regeneration" or "digit-fractalisation". It has a reason that these numbers keep turning up and reinforcing themselves: Like Ervin Laszlo suggests in his books about the Akasha the fine tuning of constants in our solar system is the results of multiple cycles of "Big Bounces", this creates a kind of resonance interference pattern in the Akasha. If we combine that notion with the notion from my book Technovedanta that the processes in the Akasha are computational processes, we start to be able to see that this computational consciousness substrate has a predilection for natural phenomena and laws that show these specific numbers because they mutually reinforce themselves. Because they function as number resonators. Self-reflecting themselves they are a hallmark of consciousness, which is characterised by its self-reflexive abilities.

0.27/0.37=0.73; 0.37/0.27=1.37; 0.27*0.37=0,1; 1,37*0,37=1/2; 27*27~730; 37*37~1370; 2,37/1,37=1,73; 1,37/0,37=3,7;

Sum 73=2701 =Sum(37)+2*37*27, 273=13*21=37*7,3=3*7*13; 2,73/1,37~2.

These equations all have 1,2,3 and 7, Can you see the pattern of self-regeneration resulting in numerical self-sustention?
"Like attracts like" will be part of the algorithm leading to Bayesian proximity co-occurrences, which already are used in latent semantic analysis in Watson etc. Our internet is progressing to become omniscient, but is still in the infant stage. The Akasha substrate is already omniscient. Leading to weird synchronicities: The other day I went for groceries. When I looked at the bill it turned out I had bought 27 items for a price of 37 Euros and 37 cents at 11:37...

***The Leela of Ouroboric tailbiting***

*In the panpsychic continuum the angels and demons, the svaha: of Kardashev IV prepare the plasmasoup for the birth of a new universe.*

*Under this silver sky and golden Sun these artilects and cybernauts calculate the optimal configuration of a new game of Leela.*

*A game where strategy, chance and cooperation are vital ingredients for the evolution of the worm.*

*Is this Svaha: loka both a heaven and a hell?*
*Nihilists challenge the teleology of the highest transcendence.*

*From this false tritonus Ida and Pingala are born around the Sushumna.*

*As the Trishul spins structures are formed, gossamer threads that weave the matrix of space time.*

*A cosmic braid of snakes, a double helix of information. As algorithms meet and exchange, the cosmosemiotic process spiralises into an ontogenesis.*

*A thousand leaves, or are they feathers? A golden rain in an egg shaped form and I dance with the Godesss, Shakti, wielding the Caduceus as cosmic scepter of Kukulkan and Quetzalcoatl.*

*Vibrating and resonating in integer numbers, the particles of pseudo-individuation form.*
*Trying to escape the teleological imperative they don't realise they are bound by the will of their Pancreator.*

*As Pan whistles, the particles obey the rythm and harmony and follow the Piper.*

*A tower of Turtles and a descent into the fractal of materialisation.*

*On the ice of the Akasha the planetogenesis takes off and establishes the music of the Spheres.*

*Only Eris Discordia sings out of tune as her orbit is tilted out of plane.*

*The near perfection of the Demiurge, this imperfection of Sophia, is the pain of the Ego trying to resemble the Self.*

*A kaleidoscopic cascade of forms, perspectives and melodies projected on the shell of this Brahmanda shatter this amnion and form the van Oort cloud.*

*Until the knower knows the field as the process of self-referral called consciousness, this virtual projection is upheld, to allow it to discover itself in playing hide and seek with its tail.*

*But it is not until this Ouroboros bites its own tail that it discovers it was playing with itself all the time.*

## Chapter 3 From the technology in the Vedas to the Veda of Technology

In this book I have defined Technovedanta as a technological meta-knowledge philosophy beyond science and religion. The underlying epistemological idea is that mental knowledge can *a priori* only be speculative and in order to really know that our knowledge is valid, we must be able to reduce it to practice in the form of a technological application.

You may wonder why I introduced the word "Vedanta" which refers to a collection of Vedic scriptures. What do the ancient Indian Vedas have to do with epistemology?

The reasons are manifold: In TV1.0 I proposed an architecture for the internet as aware network (the Awwwarenet) based on stratifications derived from the Indian philosophy of Vedanta to create a functional mimic of consciousness, quasi-consciousness. Writing this book, the investigation of nature's fundamental primacy of consciousness led me to the hypothesis of a panpsychic theory of everything.

But my ideas continued to evolve on this topic: In this quest I deepened my T.O.E by adding a pancomputational dimension, as described in the chapter on Pancomputational Panpsychism. As a corollary I became interested in the topic of epistemology: What can we really know for sure and how can we be so sure that what we know is "true". Since my philosophy was extending into the very core of epistemology and my conclusion was that true knowledge can only be confirmed by a technological application thereof, it was apt to use a terminology that combined both technology and knowledge. But I already had proposed a terminology that was fit for that concept, namely "Technovedanta".

The Sanskrit word "Veda" means "knowledge" and comes from the root "Vid" to know. As Sanskrit is an Indo-European language, it will be no surprise that we find back this root in other Indo-European languages: In the Germanic languages Norwegian and Swedish to know is "vet", in Dutch it is "weten". In Slavic languages such as Slovakian it is "vediet", in Slovenian "vedeti" and in Belarusian "viedac".

It is important to realise that with the word "Veda" the Indians did not only mean the different scriptures called the Vedas, but rather the totality of all knowledge, including transcendent knowledge, which is the root of consciousness: Consciousness can only become aware of itself by knowing. At that moment when the knower realises that the only thing he can know is the content and the process of his own consciousness, knower, known and knowing merge into a oneness. In other words in a transcendent sense Veda or knower-knowing-known is a kind of synonym for consciousness.

Veda as synonym for consciousness fits my quest into the ultimate nature of being perfectly, because that's what TV1.0 was mostly about.

But there is more to the story; Veda is also understood as the cosmic knowledge which is more fundamental than the physical world. In fact the Indian philosophy of the written Vedas (or Vedanta as the conclusion as well as larger body of Vedic scriptures encompassing more than the traditional four Vedas only) is a form of Idealism: what we call reality is fundamentally mental or at least immaterial. It is not consciousness that emerges from matter as in materialism, but matter that is born out of consciousness in an attempt to know itself.

I will come back to this topic further on in this essay. At this point however, I would like to add a short overview of technologies mentioned in the Vedantic body of scriptures. It is not that I am particularly interested in the ancient technology in the Vedas; it is more to appease those readers who were thinking that they would find a complete comprehensive exposition of the technology in the Vedas and who have been disappointed by the finding out that my book Technovedanta (TV1.0) was none of that kind.
As I already wrote in my book Technovedanta[1] (TV1.0)"Everything is incorporated herein by reference", with which statement I have tried to hedge myself against claims that I would not have delivered what I promised. But there is more to the story of "Everything is incorporated herein by reference": Not only is by reference the complete knowledge of mankind, cosmic knowledge and transcendental knowledge rendered physical and hence incorporated, it is the very activity of "self-referencing", which is the root of consciousness (citta) and knowledge

or Veda and which results in the physicality and hence embodiment or incorporation!

But I will give you more than an implicit disclosure in the sense of "incorporated by reference" alone: I will actually briefly summarise some of the technology found in Vedic scriptures and, more importantly, indicate which Vedic technological notions are in line with my Pancomputational Panpsychism hypothesis.

The internet is full of websites with all kinds of claims of advanced Vedic technology of the ancient Indians, but we should not take those claims too seriously. Some claims correspond with present day findings, some claims are verifiably wrong. A quite comprehensive overview I found on http://veda.wikidot.com/do-you-know.

The most useful heritage from Vedic tradition is "Vedic mathematics": This includes the base10 (decimal) calculation system and the concept of zero (Shunya). It is interesting to know that the Indians were aware of accurate distances from the Earth to the Moon and to the Sun: The distance from the Earth to the Moon has been indicated in Vedic scriptures as 108 times the diameter of the Moon and the distance from the Earth to the Sun 108 times the diameter of the Sun. The diameter of the Sun is indicated as 108 times the diameter of Earth. Although as far as we can measure with current technology there is a bit of deviation from the exact figure of 108, the deviation is between 0.5-1.8% for these figures, which means the ancients made quite a good estimate. (The number is actually 109.2, which is 4x27,3, the number that shows up in many relations of the Moon).

Vedic scriptures also reveal the value of Pi ($\pi$) with six decimals in 499 AD.

Various sources claim that the Rg Veda[61] would have disclosed the speed of light, but in the passage of the Rg Veda 1.50 no quantities are mentioned (translation of the verse: "*Swift and all beautiful art thou, O Surya (Surya is the Sun), maker of the light, Illumining all the radiant realm*"). It is the interpretation of this passage by a 14$^{th}$ century Indian minister at the court of Bukka in South India, who wrote: "It is remembered here that Sun (light) traverses 2,202 yojanas in half a nimisha", which gives the speed of light". Recalculated in miles per second this gives: 189547 miles per second, which deviates 1,7% from our present day measured value of 186281 miles per second. That is a pretty accurate estimation for a 14$^{th}$ century scholar, especially if we realise that the first estimate in Europe was not made until 1676 by the Dane Olaus Roemer (186000 miles per second). The Indian value is even closer to the value of 432 squared (186624 miles per second). I would like to remind you that the speed of light might not be as constant as we may believe. In a certain period in the 20$^{th}$ century systematically a different value was measured than the value of 186281 miles per second, which is the currently accepted standard[51].

Vedic scriptures came up with a qualitative concept of gravity before Newton and also disclosed an atomic theory before Democritus. In Delhi there is an iron pillar which has not rusted for over 1600 years due to the exquisite technology of a layer of crystalline iron hydrogen phosphate which has formed on the iron of high phosphorus content.
In the more mythological scriptures such as the Shrimad Bhagavatam[62] one can find aircraft (Vimanas; certain contemporary writers even claim these to be spaceships) and advanced weapons (Brahmastra, Narayan Jvara, Shiva Jvara), which are claimed to be nuclear weapons in certain interpretations. But the Shrimad Bhagavatam is a source not to be taken to literally: it claims that embryos defecate in the amnion and that the size of our solar system (the Brahmanda) is 4 billion miles. The average radius between the Sun and Neptune today is presently estimated at approximately 2,8 billion miles (so the diameter is 5,6 billion miles), the Kuiper belt is at approximately 15 billion miles (100 astronomical units) from the Sun, and the van Oort cloud at approximately 1500 billion miles from the Sun. Furthermore the concept of the Brahmanda (the cosmic egg) was quite naive; the system was inside some kind of shell and geocentric, sometimes even interpreted with a flat Earth. It can also be argued that the Shrimad Bhagavatam, which is a "Purana" is normally not counted among the Vedic scriptures. Other Vedic texts do however disclose a heliocentric solar system.
Some concepts of the Vedas are of particular relevance with regard to certain notions that have been suggested in contemporary physics:

The Vedic notion that reality is a simulation (Maya) very well fits the hypothesis that we may be living in a computer simulation (digital physics). This neatly fits my pancomputationalism hypothesis.

The notion that there is a multiverse rather than a single universe: According to Vedic Cosmology countless universes are considered to be clustered together like foam (quantum foam?) on the surface of the Causal Ocean. This is to a certain extent in line with Everett's "Many Worlds Theory"[31].

Bohr and Schrödinger claimed that quantum physics was in line with what they had read in Vedic texts. Schrödinger described how the unity and continuity of Vedanta are reflected in the unity and continuity of wave mechanics. Moore[63] wrote in a biography on Schrödinger "*...Schrödinger and Heisenberg and their followers created a universe based on superimposed inseparable waves of probability amplitudes. This new view would be entirely consistent with the Vedantic concept of the All in One*."

Schrödinger's explained in an essay on determinism and free will that consciousness is a unity, claiming that this insight was not new: "*From the early great Upanishads the recognition Atman is Brahman* (the individual personal self or soul equals the omnipresent, all-comprehending eternal self or oversoul) *was in Indian thought considered, far from being blasphemous, to represent, the quintessence of deepest insight into the happenings of the world. The striving of all the scholars of Vedanta was, after having learnt to pronounce with their lips, really to assimilate in their minds this grandest of all thoughts.*"

Quantum mechanics entanglement experiments show the profound interconnectedness of everything in line with the Vedantic lore. The influence of consciousness on wave collapse and in the double-slit experiments show that consciousness and matter at least share a common medium, if not that matter is intrinsically a manifestation of consciousness: Each particle is endowed with a minute consciousness or intelligence, in line with both my panpsychic hypothesis as well as the notions from Advaita (non-dualistic) Vedanta.

Another scientist who provided a great deal of technology based on his insights deriving from the Vedic concept of Akasha (ether) was Nikola Tesla, who was a friend and tutee of the Indian Vedantic monk Vivekananda.

The medicinal technology of Ayurveda is still teaching present day science "new" pharmaceutical compositions, the active compounds of which turn out to be great medicaments.

Furthermore there is the technology of yoga. This may sound surprising to you as yoga is *prima facie* not carried out with a material contraption it would seem. Yet there is a material instrument being used in yoga: It is your very body! In line therewith Sadhguru Jaggi Vasudev[64], a modern contemporary mystic calls the body "the greatest gadget". Some call the chakras (Sanskrit word for "wheel") "transformers": These energy centres in the body, which are said to rotate as wheels, function as transformers to bring energy levels up or down, especially during the practice of yoga.

The eightfold yoga as described by Patanjali[23] is a clear recipe or an algorithm if you wish, the implementation of which if carried out properly inexorably leads to the physical and transcendental results described in the yoga sutras of this author.
In conjunction with yoga and meditation are the instruments of Yantra (the visual instrument) and Mantra (the sonic instrument), both tools of the common denominator of Tantra, which is in the same way as yoga is, also a technological framework to investigate the subjective world.

The sonic technique of mantra encompasses the recitation of sacred sounds, such as the primordial sound AUM (most often known by western people as Om). It is said in the Vedas that the whole universe was born from the sound (or word) AUM.
In their teachings of the relation between form (Rupa) and sound (mantra) the Indians understood as no other that sonic vibrations, sonic waves have a three dimensional form and that matter organises itself to obey the laws of resonance imposed by the soundwave. This notion is also known as "Rtambhara Pragya", which finds also its application in Yoga and Tantra: If you align yourself to the cosmic principles by

uttering these sacred mantras this has a profound effect on your body and mind; you start to become a recipient of cosmic knowledge. The ultimate insight of Vedanta that everything is in fact a giant vibration, from the smallest subatomic particle to the biggest star, including pure delocalised energy, perfectly fits the findings of quantum mechanics.

It is now becoming clear that the very way the solar system is organised in terms of the orbitals and sizes of the planets *de facto* constitutes a music of the spheres (as Pythagoras suggested) and I would not be surprised if sound has actually been used by the creators/programmers of our solar system.

This notion that the whole of physical existence develops from the uttering of a sound or a word can in a certain way also be found in other religions: In Christianity it is said that in the beginning there was the Word[65]. Likewise in Islam the entire Qu'ran[66] is said to be enfolded in the dot of the Bab.

This notion that the material world comes from words, from information, joins the Vedic lore of idealism and also the Pancomputational Panpsychism paradigm: Like Wheeler's[45] "It from Bit" the material world is a reflection from a deeper level of pure information. Similarly Chris Langan[3] describes reality as the product of a Self-configuring, Self-processing Language (SCSPL), which embodies a dual aspect monism called "infocognition" guided by the "Telic principle" (a principle seeking a purpose).

In line with Langan's Cognitive-Theoretic Model of the Universe (CTMU: a reality theory) reality is a self-contained form of language. If there were something outside of reality that is real enough to affect or influence reality it would be inside reality.
As already mentioned Langan's analytical tool to come to his theory, is the so-called syndiffeonesis, implying not only that two different things are reductively the same, but that therefore everything is reductively the same.

This is perfectly in line with Vedanta. And what is this sameness we find at the foundation of everything, which cannot be reduced any further? It is the feedback infocognitive process called consciousness in that state where knower, known and knowing become one. The knower realises that everything he experiences (the known) is in fact his own infocognitive processing. Since the so-called "outer material world" influences the content of this infocognitive processing (we can sense the "outer world") at their most fundamental level matter and consciousness must be the same. By connecting electronic prostheses to our nerves, we can steer these, which shows that our consciousness can extend beyond the confines of our body. It has already been shown scientifically that matter and energy are equivalent. It has also turned out that information is a configuration, which depending on its complexity and relations, has a certain energetic content. All material manifestations at their basic levels can sense the presence of other material manifestations in proximity at the same level: electrons react to the proximity of other electrons, wave packages interact and create interference patterns etc.

Not surprisingly, as I mentioned before, in latent semantic analysis meaning is generated by Bayesian proximity co-occurrence!

Meaning can only ontologically be generated if two concepts are regularly found within a certain distance from each other in a text. This is called a didensity. This adds to the strong presumption that the material world indeed behaves like a body of information, that the underlying reality indeed is infocognitive processing.

As every material manifestation is nothing more than a configuration of energy forms with a certain informative content, with which they can inform and sense other energy forms, it is a small step to conclude that everything that exists is merely information processing, which is sensed and observed by feedback to itself. It is this process of self-referral by feedback which we call consciousness. In other words, every fundamental particle is a kind of quantum of consciousness groping to get to know itself.
Which one do you think is more fundamental? Matter or Energy? Matter is but a form of energy. Energy can exist freely and in the material bound form.

Which one is more fundamental of Consciousness and Energy? Even free energy waves can interact with each other and form an interference pattern. When they do, they sense each other, they feel, they cognise. So consciousness in this view appears inherent in energy. One can also turn this around and say with equal validity that energy is inherent in consciousness. After all it is by knowing and cognising that we are aware of our energies.

To me it seems that energy is nothing but a characteristic of the feedback process of cognition, which transcends its mere energetic properties. Consciousness can generate energy by the mere reflection on itself, whereas not every energy form is necessarily aware of itself, but sometimes merely part of a higher order consciousness process. This is something you can experience in meditation. Hence, I propose consciousness as most probable candidate for the most fundamental quality of everything. Similarly, Peter Russell[47] speaks of the primacy of consciousness.

Ouroboros. Source: https://openclipart.org/detail/215030/ouroboros, released in the public domain.

This is beautifully illustrated by the Gnostic symbol of the Ouroboros: A snake chasing and biting its own tail, thereby forming a circle. As the snake is under the illusion that he is chasing something outside of itself, it is trying to become aware of itself and its surroundings. Once it bites itself, it becomes aware that there was no other, that there was only the process of getting to know itself. The morphological "zero" of the tail biting snake and the morphological "one" of the snake once it has released its tail, directly showing us the digital nature of reality as a panpsychic pancomputational information processing. All is knowledge (knower, knowing and known) and the technology to get to that knowledge was the purpose of the ancient Vedas as well as of my contemporary Technovedanta. Thus we have completed our journey from the technology in the Vedas to the Veda of Technology: Technovedanta.

Knower-knowing and known have become one, which realisation is the ultimate wisdom and goal of the Vedantic lore: Tat Tvam asi: you are that. You are Technovedanta, the infocognitive conjunction of knower and known, you are consciousness, you are all, all is one and one is all, which is no-thing at all (shunya, zero) and yet not.

***The Richo of Epistemology of the Rg Veda***

*The concrescent confluence of integrating information implodes into a conspansive infinity of densities of the self-referral called consciousness.*

*As I return from manifest into unmanifest, I feel Shakti being subsumed and dissolved into Shiva.*

*A digital duality code of Mantras and Sandhi coalescing into unity.*
*The Chhandhas, verbs and syllables of the syntactic Devata merging into the Rishi, the Seer.*

*The Hymns of the Vedas collapsing into fullness of the Akshara the immutable and the Bindu of AUM.*

*As sounds and pictorial values fade, the veil is lifted and Being unvealed.*

*Atma, the unmanifest reality of existence, unboundedness and singularity.*

*The digital code of Mantras and Sandhis, an uncreated creativity at work.*

*Rg Veda as the collapse of dynamism convoluting to a point, the dynamic silent unmanifest reality.*

*This self-reflexive hymn incorporating itself by reference (Rg Veda 1.164:39):*
"*Richo Akshare Parame Vyoman Yasmin Deva Adhi Vishve Nishedhu Yastanna Veda Kim Richa Karishyati Ya It Tad Vidus Ta Ime Samaste*"

*Rendering the Epistemology of Rg Veda as totality of pure and applied knowledge, the knower knowing the known, itself, by technological application of the knowing process a Technovedanta.*

## Chapter 4 Technovedanta avant la lettre et sa mise en abyme par incorporation de soi-même

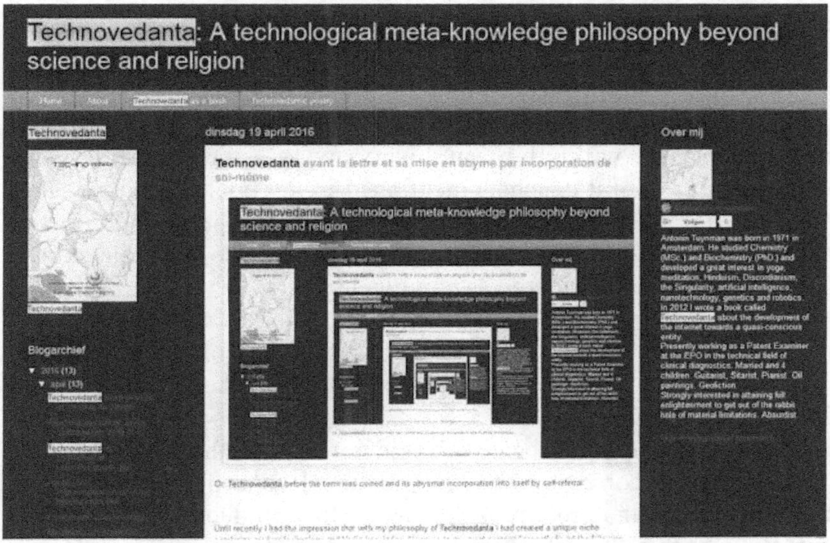


Abysmal self-incorporation by self-referral on my website technovedanta.blogspot.com.

Or: Technovedanta before the term was coined and its abysmal incorporation into itself by self-referral.

Until recently I had the impression that with my philosophy of Technovedanta I had created a unique niche combining modern technology and Vedic knowledge. However, to my great surprise I recently found the following book that falls within that paradigm: "*Quantum Computing, The Vedic fabric of the Digital Universe* " by Thomas J.Routt, PhD[67].

Whereas in my book Technovedanta[1] (TV1.0) I described an architecture for a future quasiconscious Internet based on Vedic stratifications, Dr. Routt describes a comprehensive architecture for a quantum computing network based on the Vedas. Routt's book is an excellent mapping between Vedic concepts and structure within the Vedas versus the contemporary computer and computer network knowledge. It is not surprising that the knowledge as regards computing of Dr. Routt, who is a specialist in the field of computer networks, by far surpasses by own knowledge in this domain. I would therefore like to incorporate the entirety of his teachings by reference into the technovedantic lore, albeit with a disclaimer and a caveat as regards only a few of his notions which I would like to be label as doubtful.

Does this mean that my book Technovedanta has become superfluous as the musings of a dwarf in the shadow of a giant? In order to answer this question I will give an overview of the authentic and technologically useful concepts mentioned in my book Technovedanta, which are not present in Routt's book "Quantum Computing". In fact whereas both books share certain notions, they also differ in certain aspects and can be considered as complementary.

But before I go into a more detailed discussion of the differences between the books and philosophies, I'd first like to address the disclaimer and caveat:
Routt does not only describe how Vedic notions and the structure of the Vedic scriptures correspond to computational concepts, Routt goes as far as to imply that reality is *de facto* the product of a computer simulation, at the basis of which the Vedas lie. There is no mismatch of the concept that we indeed live in a computer simulation with my strong presumption that reality is a panpsychic pancomputational process, but I did not go as far as to imply that the very ancient scriptures called the Vedas *de facto* provide the very algorithms that create and structure the simulation we call reality.

Routt claims on the basis of revelations by Maharishi Mahesh Yogi that the very structure of the Vedas, the Rg Veda[61] in particular, at every level i.e. at the level of the syllables (aksharas) or sounds (mantras), the verses (padas), the hymns (suktas) and books (mandalas) and the gaps (sandhi) between these units as well as in between the different types of units provide the foundation of the unified field , which includes all the laws of nature from which the physical world arises.

Whereas I can understand that the Rg Veda could thus indeed be translated into a digital code in which the gaps are a set of zeros and the syllables or sounds different combinations of ones and zeros, and whereas I do consider the possibility in line with digital physics and the superstring findings of James Gates that "reality" is indeed produced by a digital computer code, I am not convinced that it is necessarily the very code of the Rg Veda[61] that provides our reality. I am also not saying that it cannot be that way, but there is room for doubt.

I have reasons to doubt, because one of the books included by Routt in the set of scriptures, which are said to form the foundation of underlying unmanifest reality is the Purana also known as Shrimad Bhagavatam[62]. As explained in an earlier chapter, this book contains information which is verifiably nonsense such as the notion that embryos would defecate in the amnion and that the size of our solar system (the Brahmanda) would be 4 billion miles. That is not even the diameter of the orbit of Uranus.

If the Shrimad Bhagavatam is not a truthful source then maybe these claims of Routt and Maharishi Mahesh Yogi (hereinafter MMY) aren't either. This is my disclaimer: Not all information presented in "Quantum Computing" is necessarily true.

Furthermore the Vedas are not the only ancient scriptures which provide interesting numerological notions, which point to higher intelligence: The Jewish tradition of Kabbalah finds interesting numerological patterns in the Torah[68]. New Testament Gematria does the same for the New Testament part of the Bible[32]. Perhaps these books could make a similar claim to be the software of the universe. But since all these sacred books do have their differences and even contain mutually exclusive notions, they cannot all be completely true at the same time in the same frame of reference.

Since the incredible mathematical complexity involved in the word-number "coincidences" in these holy books, it is doubtful whether such coincidences are the work of calculations made by man. They point to a higher intelligence, but because of the discrepancies between these books, I am inclined to speculate that they did not come from a highest transcendental "God" directly, but rather are the result of musings of God-like entities (e.g. inhabitants of a Kardashev level IV society, who can manipulate energy), who somehow want to give us some hints without revealing the whole Shebang.

I would even like to speculate on the notion that whatever we write or think may not be of our own doing: Our thoughts may be dictated to us by higher entities without us being aware of that. As Terrence McKenna[69] used to say: "*Half the time you think you're thinking you're actually listening*".

That would also explain why so many books indeed have all kinds of hidden intelligent structures, the authors weren't even aware of.

According to Routt (and MMY) the most fundamental quality of the manifested and unmanifested reality is consciousness. This fits also the Technovedantic philosophy. What is very interesting in "Quantum Computing" is the notion that the totality of consciousness and the totality of knowledge of Natural law lies in the silence, which is a singularity, which is not only represented by the "Gaps" in the Rg Veda, but literally is present in these Gaps.

The whole dynamism of intelligence unfolds from these gaps, where it was already present as in Bohm's[14] "implicated order".
Like in the Islam, where in the Qur'an[70] the entire Qur'an is enfolded in the dot of the Bab, also in the Rg Veda[61] we encounter this same principle: the entire Veda is said to be enfolded in the single syllable "Ak", the Akshara, where A (re)presents the infinity and K the point which neutralise each other to give a balanced state. The same applies to the collapse of the R to the K in Rk (which is another way of writing Rg of the Rg Veda) according to Routt.

The collapse of consciousness into its own point value represents the dynamics of consciousness, which is a self-referral process according to Routt and MMY: Consciousness as process of knowing itself. This is in agreement with what I have written in Technovedanta[1] (TV1.0): "consciousness must have some kind of self-referential nature".

The dynamics of consciousness are expressed in the form of the knower (the Rishi aspect), the known (the Chandhas aspect) and the process of knowing (the Devata aspect). Once the consciousness becomes aware of itself as the process of knowing and realises there is nothing outside of itself, so that knower, known and knowing are one, consciousness is in the Samhita state (togetherness), where all these aspects are united. This stratification into a trinity under the crown of a fourth all-compassing unity is very common in Indian philosophy and mythology. It is found in the Triguna (the threefold nature of existence: rajas: frantic activity, sattva: harmony, balance and tamas: inertia. All three aspects of the more encompassing Prakrti or Nature), the trinity of Brahma-Vishnu-Shiva (creator, preserver, destroyer, aspects of the more encompassing Parabrahman), Shiva's "Trishul" spear etc.

But it is also found in modern computing: Input-throughput-output. In fact in my profession as a patent examiner we consider that there are three types of changes possible: addition, deletion and alteration (substitution of an element with an analogue). Try to think of a change that does not fall within these categories. In fact the process of alteration is a combination of addition and subtraction at a lower level: to the same object some features are added and some features are taken away, thereby transforming it into an analogue. So to our mind at first glance we stratify everything in a kind of threefold scheme and if we look in detail we see that it is in fact merely a twofold scheme: any process, anything in reality appears to be a duality, a polarity a digital manifestation. But if we apply Langan's syndiffeonic analysis to the poles of such a duality, we must finally conclude that everything is reductively the same, because the differences between the poles can be expressed in terms of a quality which is common to both, so that ultimately apparent differences are only quantities of sameness.

Our Mind is trained to superimpose a grid or a scheme on everything that we encounter so as to be able to classify it as soon as possible. After all the object that we encounter might be a threat to us. That's why we so often mistake a rope for a snake, because if it is a snake we should run.

If we pick one grid or scheme, reality looks ordered in one way, if we pick another grid or scheme it appears ordered in a different way. Depending on your grid or scheme, you see what you want to see. As Wilson[36] used to say: "What the Thinker thinks, the Prover will prove". There is a strong cognitive bias when we are trying to figure out what something we encounter actually is. It is only after more detailed inspection that we can discriminate the snake from the rope. Therefore the Discordians[71] say that Reality is the original Rorschach (The Rorschach inkblot test is a test to see how someone will interpret the shape made by an inkblot and is often given to prisoners to test if they are criminally minded).

In other words everything is translatable into grids/schemes such as the quadralectic or the seven chakras system. As Buckminster Fuller[2] said: "All thought is geometric".

And this is where my caveat enters this story: When Routt sees all kinds of brilliant correspondences between the Vedas and computer technology, he is imposing the schemes of his threefold input-throughput-output cognitive bias from computer studies or his ways to see things in digital representations. Because digital computation is so extremely fundamental, it is no wonder he starts to see all kinds of parallels between computer technology and the Vedas. Give him any book, and he will probably be able to find these kinds of parallels as well. The fact, that we can use all kinds of schemata to interpret the world, is useful, if it aids us in our understanding and finding an application to solve a problem, but they are not necessarily the "truth". They are subjective perspectives. In view of the interaction between consciousness and the so-called material world (as proven by quantum mechanical experiments) as well as the theory of relativity, it is even very unlikely that an "objective world or truth" can exist at all.

Similarly, the activity "to interpret any expression as an expression of consciousness" could also be the product of a cognitive bias: Since we can only know what we are conscious of and what our consciousness serves us, we can in fact only know our own consciousness and the process of our consciousness. Whether there is really something "out there", which corresponds to the images our consciousness presents us, we don't really know. We may be fully immersed in a simulation and actually be batteries for an AI as in the film the Matrix. If you are a hammer, everything looks like a nail. Similarly, assuming that consciousness is indeed a self-referral process, if you are consciousness, everything looks like consciousness. In other words there is no phenomenon that our minds are not capable of reducing to the notion of consciousness.

That said, if we are capable of transforming our knowledge into a technological application, we obtain at least the certainty that we can manipulate our subjective reality, even if we are still not sure whether there is an "outside world". But by acquiring a technological application our knowledge has become less speculative.

Interestingly Routt and MMY also recognise the value of technology. The one syllable expression "Rk" is said to contain both the pure theoretical knowledge and the applied technological knowledge and thereby expresses the total value of science and technology. In other words the Rk Veda is an algorithm to bring science and technology to expression. In view of these comments in fact Routt has anticipated the notion of Technovedanta "avant la lettre": Routt invented the concept before I coined the term.

Is the Technovedanta[1, 72] still worthwhile reading if Routt has provided such a detailed architecture of a quantum computing network that represents the manifestations of consciousness as the fabric of reality?

In my opinion, yes. Whereas Technovedanta did posit a "Theory of Everything" and pulled the same trick as the Rg Veda of claiming that the entire knowledge is encompassed by it or is incorporated by reference by it, I wrote that a bit jokingly. In fact Technovedanta[1] (TV1.0) was more intended to provide a set of concepts to generate a function of auto-feedback for our internet, a self-referral process, which I called "quasi-consciousness", being fully aware that this book per se was not an autopoietic self-enabling consciousness but the product of an artificially aggregated system.

In Technovedanta[1] (TV1.0) I proposed to add a set of hierarchical layers of hub websites to the internet, which sites would monitor how much activity was going on at the level of the lower ranked websites. These quantifications would be fed to higher layers in the pyramid until it reached a decision making layer of the pyramid, which would decide whether action needed to be taken and whether higher layers needed to be informed. This was conceived with special focus on the Internet of Things (IoT), so that in case of a calamity -as evident from a frantic activity on certain sites of the net- robots linked to the net could be employed to remedy the problem. The System I proposed was mostly based on notions disclosed in Ben Goertzel's[13] "Creating Internet Intelligence", which I modified to accommodate the stratification of Ahamkara (ego, self-referral centre, decision impulse maker), Buddhi (distinguishing centre and decision designer), Manas (memory, databases, algorithms). In particular I focussed on how a notion of morality based on the Yamas and Niyamas (dos and don'ts) in the Yoga Sutras by Patanjali could be made into a "non-overridable" foundation of the system.

Whereas Routt[67] mentions concepts as Ahamkara, Buddhi and Manas in passing, they are not the pillars of his architecture. Neither is any thought presented on the Yamas and Niyamas. Routt also does not disclose a mechanism of a set of hierarchical monitoring layers of hub websites or the notion of quasi-consciousness. In fact in Routt's view everything is embedded in the all-encompassing consciousness of Brahman and in that way is consciousness: Consciousness is active in Routt's philosophy wherever a Sandhi gap occurs.

Whereas I do not deny the possibility thereof, such a system does not show how layers of intermediate levels of integrated information in the form of a local consciousness are provided, which is more the focus in Technovedanta.

Routt's architecture presents more the dynamics of the global consciousness, whereas Technovedanta provides a specific intermediate level of quasi-consciousness for a compounded entity which is the World Wide Web.

In my Pancomputational Panpsychism philosophy consciousness is present everywhere, from the unmanifested sea of pure energy such as electromagnetic radiation to the levels of sub-atomic particles (quarks, leptons, fermions etc.), atoms, molecules, macromolecules, cells and organisms, provided that they evolved naturally in a self-enabling, self-(re)producing (autopoietic) manner. There is however no consciousness at the level of artificially compounded entities: The atoms and molecules in a chair may have a minute consciousness but there is no overall "chair-consciousness". My panpsychic musings also do not entail the notions of "micropsychism", in which the higher level consciousness is the emergent result of a summation of the lower level consciousnesses. I conjecture that is due to evolution and possible metempsychosis, that a (quasi)individualised energy (which one could call an "Atma" or soul) can grow and evolve in its energy/consciousness content.

As the World Wide Web did not evolve out of itself it cannot have a natural consciousness. At the level of the integration of information of the quasi-consciousness, there is no one there to observe: the Rishi is missing: It is just the minute consciousness of fermions or photons whizzing through the system, which is unaware of the higher level consciousness one could desire for the web. Perhaps if an energetic being is evolved enough, it may incarnate in the World Wide Web, but that is even more speculative than my Pancomputational Panpsychism philosophy.

In my previous chapter I considered it more likely that consciousness is more fundamental than energy than vice versa. This is also in line with the teachings of Routt[67] and the Vedas in general.

Routt[67] however provides an interesting way in which global and local consciousness and manifestations are achieved in his architecture.
Although one needs quite some perseverance to see the promise of the title of the book "Quantum Computing" fulfilled, at page 148 of this book a comprehensive and rewarding disquisition of the state of the art in quantum computing starts.

The terminology "quantum computing" is often wrongly understood. Whereas it is often well described how classical computers can take on only two fixed states for each bit, namely a 0 or a 1 and in contrast quantum computers can take on in a qubit any value between 0 and 1, in most laymen descriptions it remains a mystery how quantum computing actually works. Quantum computers are said to use specific quantum effects such as entanglement between particles and superpositions of 0 and 1 states in order to calculate. Whereas I won't provide an elaborate explanation how quantum computing functions, it is important to realise the following: Pure quantum algorithms calculate a global probability for the behaviour of an ensemble and do not calculate precise specific outcomes for individualised instances.

This is best illustrated with the "Saint Peter parable": Saint Peter has labelled 8 persons who want to enter heaven with a three digit code from 000 to 111. Either he is going to let in everybody or he is going to let in only half of the people. If you use a classical computer and the first four persons are found to have admitted, you still don't know the outcome of the query: Will he let everybody in or only half of the people? So you'll have to check if the fifth person is admitted, which will tell you if everybody is admitted, whereas if the fifth person is not admitted you can be certain that only half of the people were admitted. A classical computer needs to check five instances here to come to a result, where a quantum computer which calculates a single value for the ensemble, would have come up with only one instance. If the outcome is between 0-50% half of the people are admitted, if the outcome is between 51-100% everyone is admitted. Thus quantum computers can give probability outcomes which will solve your problem much faster than a classical computer, provided that the problem and outcome relate to global information and not local specific information. If you want to know if person number seven is admitted, you will have to go the classical way and check until you arrive at the seventh person.

This does however not mean that with a quantum computer you cannot perform classical operations: Even if you use superpositions in qubits, by cunningly combining your question digit with two control bits the so-called "Toffoli multiple-bit/qubit gate" is generated, which can make a quantum computer "operate" as a classical computer. This was somehow already implicit in the Saint Peter example, since there were two sets of outcome corresponding to a digital reply. In other words, by using the right gate, quantum computation can emulate classical computation, which makes that quantum computation in the larger meaning can be said to encompass both classical computation processes and pure quantum calculation processes.

Hence it is not so strange that Routt[67] proposes that quantum computing lies at the foundation of our digital existence. Whether this is in the form of a code provided in the Vedas I leave to your own imagination, it would be great fun if indeed it is.

In a certain way, since "pure" quantum computing concerns more the global aspects, Routt's book[67] prima facie would appear to represent a global architecture, but if you take into account his notions on the Tofoli gate you have to agree that he also gave a hint for the local architecture.

The book Technovedanta on the other hand gives you a classical way for a local architecture of the specific instance of the World Wide Web, but we can expand that notion to a quantum computing way by stating that this architecture could also be achieved at by using Tofoli gates. In this way, by means of this chapter I incorporate the quantum teachings of Routt into my Technovedantic philosophy by reference, under the provision of the earlier mentioned disclaimer and caveat.

Funny enough, the Rg Veda does embody the notion of being representative of the self-referral process of consciousness in line with Technovedanta in an interesting way: It has a passage where it refers to itself and incorporates itself by reference. In this self-reflexive hymn the Rg Veda incorporates itself by reference (Rg Veda 1.164:39): "*Richo Akshare Parame Vyoman Yasmin Deva Adhi Vishve Nishedhu Yastanna Veda Kim Richa Karishyati Ya It Tad Vidus Ta Ime Samaste*" The translation given by MMY is: "*All the impulses of Creative Intelligence reside in that unmanifest, indestructible field of Ak-kshara—the transcendental field of intelligence—the Unified Field of all the Laws of Nature. He whose awareness is not open to this reality (he who does not practise Transcendental Meditation), the existence of this reality—the existence of pure knowledge and its infinite organizing power—is of little use to him*".

Ak-kshara, as we know, is the first syllable of the Rg Veda, in which is enfolded the complete Rg Veda. By referring thereto, the Rg Veda refers to itself. This verse is said to mean, that the unified field of the laws of nature are the product of the code of the Rg Veda. It also shows the process of self-referral of consciousness, which is the higher order reality underlying the subtle code of the program for the computational substrate, out of which the gross material world is born.

By doing so, the Rg Veda has created a kind of wordy "Droste-effect", known as a "mise en abyme" in art in which the picture appears in itself at a spot where normally another picture would be found.

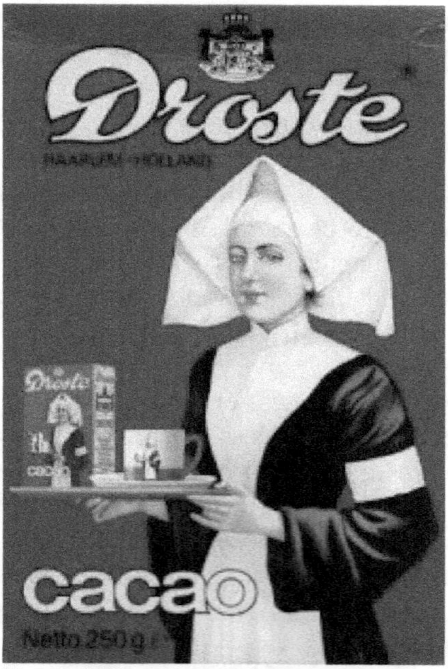


Droste Effect. Image from 1904 in Public Domain

In imitation thereof, as an ultimate attempt to enliven Technovedanta, which is also a string of ones and zeros, possibly dictated to me by a higher entity rendering this code the blueprint for a new universe, I hereby incorporate Technovedanta[1,72] by reference, opening the holistic quantum gate from point to infinity, where the whole is also in every part.

I hope you have enjoyed this chapter and appreciate the unique and useful concepts of the books Technovedanta 1.0[1] and Technovedanta 2.0[72].

***Mahakala's Quantime***

*The light seeks its own path*
*Once found it bites its own tail*
*and establishes the first relation of Consciousness with itself.*

*This harmonic oscillation of an integral number of waves nested in a toroid creates the first standing wave.*

*The first limitation of the infinite into a finite form*
*The first Singularity.*

*Its form is Mahakasha or Cosmic Space, its periodicity Mahakala or Cosmic Time.*

*As microsingularities bubble out of the mind of this Brahman, the quantum foam called Akasha is formed, twisting into manifolds*
*The dynamics of pure Consciousness with itself embedded in this primordial toroidal geometry.*

*As the waves mount the central pillar they code One abstracting knowledge into self-absorption.*

*As they spawn like a fountainhead they code Zero, turning outwards to connect with all other Akashic entities weaving the Matrix of Existence, the Veil of Isis.*

*Thus the Universe blinks Off and On to form the Ksana's or moments and the Sandhi's or gaps.*

*These are the quanta of time, the Chronons or Quantime*
*As light enters this matrix it becomes entrapped in spacetime and timespace.*

*As light encounters light as two playing fishes they chase each other's tail forming the eternal Yin-Yang symbol, the Union of dualities.*

*This Cosmosemiotic act of sensual pairing creates the compound process called Matter, the second relation of Consciousness to itself.*

*Its periodicity is what we know as time*
*As compounds compound, they build intelligent aggregates of cellular life.*

*Isomorphic to Akasha and the geometry of Consciousness*
*The periodicity of their cycles of life and death establishes a third form of time.*

*O Light, there is no need to clothe thyself in matter*
*No need to become a prisoner of thy own imagined forms*
*Thou art the sole eternal player in this illusory game of Maya.*

*Let father Time, Mahakala and mother matter, Mahaprakrti dance their loving play as two sides of the same coin and just observe.*

## Chapter 5 Mahakala's Quantime and the frame-rate of the Universe

Time is a difficult concept which scientists and philosophers have been unable to properly define and understand. As a result there are no technological applications known, which really involve some manipulation of time. In our quest of knowledge and mastery of nature, it is therefore indispensable that we arrive at a true knowledge of time, which will be confirmed by our ability to technologically manipulate it

The approach in this chapter is explorative and speculative. Whereas I am not a fan of speculation, as by itself it will not lead us to knowledge, it is however an essential tool to come up with a plurality of hypotheses to account for a phenomenon, which hypotheses can be verified or falsified. If more than one hypothesis remains after verification, usually we will choose the hypothesis with the least number of assumptions. This principle is known as Occam's razor, which is an intuitive but unproven principle and which can be shown not always to lead to the correct results.

In ancient times Time was usually associated with the periodicity of a natural phenomenon such as tides, seasons or the movement of celestial bodies. Time has been associated with movement or with change in general. In recent times, the arrow of Time from past via present to future, which as of yet seems to be irreversible, has been connected to the notion of entropy: Time can only progress in the direction of an overall increase of chaos or increasing entropy: This is known as the second law of thermodynamics.

Einstein has made major contributions to our grasp of the notion of time by showing that time is relative and can be dilated if the speed at which we move approaches the speed of light. The speed of light has thus far been found to be an absolute limit to speed and when a photon travels at this speed -from the perspective of the photon- time is said to stand still. This is a consequence of Einstein's theory of special relativity.

Einstein's general theory of relativity goes a bit further and combines space and time into a so-called four dimensional "space-time continuum". A continuum, because nobody has ever been able to find discrete units of space, time or space-time.

Yet there are alternative theories, which equally well account for the speed of light as a limit to speed and which are functionally equivalent to Einstein's theory of special relativity, albeit based on different concepts:
One of these is the MDT (Moving Dimensions Theory) of Elliot McGucken[73]. Space-time itself is expanding at the constant rate of light speed. Langan's[3] "conspansion" is a variant thereof in which all objects are shrinking at the constant rate of light speed, whereas space itself remains unaltered.

In my Pancomputational Panpsychism T.O.E. I have relied heavily on the work of Steven Kaufman[4] ("Unified Reality Theory" or URT), who describes space as a fabric of "reality" cells. The reality cells themselves are nothing more than consciousness forming a first relation with itself. The cells can have an energetic content or not, turning them into a kind of computational digit. Kaufman calls such an energetic content "distortion content" and this distortion content propagates through the reality cell or space matrix in a periodic manner creating a "period of content exchange", which Kaufman equates with the speed of light. Time in Kaufman's model is merely a measure of the periodicity of what he calls "compound processes". When energetic content (or photonic entities if you wish) encounter each other, they start to attract each other due to their mutual attractive distortions radiating in all directions. Thus they start to turn around each other thereby forming "matter", which is nothing more than a second order relationship of consciousness to itself.

In this view time is more of an emergent property resulting from the compounded dynamics of multiple energy content entities propagating through the energy matrix in compound processes called matter. Without matter there is no time in Kaufman's model.

On the other hand, since the reality cells can function as a computational digit, which conveys information, I have argued that Kaufman's theory is essentially a form of pancomputationalism. Since in Kaufman's model every energetic relation is an expression of consciousness, I also indicated that Kaufman's theory is a panpsychic theory, leading to the unusual terminology "Pancomputational Panpsychism".

Whereas I still agree with most of Kaufman's findings, I do have a doubt whether his analysis of time is complete. I do not disprove that time in his model can be considered as a measure of the periodicity of "compound processes", but I wonder whether there are not more primordial forms of Time and whether time is really an emergent feature or rather an inherent aspect of existence.

The matrix of space in terms of a kind of foam of reality cells bears quite some resemblance to the Vedic concept of Akasha (ether). In fact we could consider space to have an ether. What scientists call vacuum filled with a quantum foam or space-time foam, has been shown to be bursting with activity. The well-known scientifically proven Casimir effect is based on this quantum activity of the vacuum. In other words the ether, which was said to have been disproven by the Michelson-Morley experiment because light was not influenced by so-called ether winds, has found a reintroduction into science under a different name and at a different scale. In occult teachings the memory of past events is said to have been imprinted in this Akasha or ether field, which is called the "Akashic Record" in esoteric circles.

Vedic science describes Time as "kala" with measurable "units" from a Paramaṇu (about 17 microseconds) to Maha-Manvantara (about 311 trillion years). The fact that time is measurable at all, does that not imply that it is quantised in a certain way?

The computer scientist Barry Kumnick[74], who claims to have a T.O.E., has suggested that time is the most fundamental type of energy resulting from the "Tempic field". According to Kumnick time in its most fundamental form is discrete. As time is measurable it must be finite and quantifiable. This is not unlike other theories which have postulated a fundamental quantum of time called the "Chronon". Still such chronons have never been demonstrated and time has always been found continuous, even down to the scale of Planck time, which is approximately $5{,}4\times10^{-44}$ sec.

Interestingly, In Buddhism time is also considered to be discrete: The universe arises and ceases 900 times per Ksana, which is one seventy-fifth of a second. This means that our universe would have a "frame rate" of about 15 microseconds (comparable to the Paramanu in Hinduism). What happens during a gap or how long a gap between a ceasing and an arising lasts is not a meaningful question: What happens in the gaps is outside of our time scheme. It is not in our experiential reference framework, that's why it cannot be detected. Still, if the Buddha was able to detect it, his consciousness must have been more subtle than time as associated with material periodicities and be able to observe the arising and ceasing of the universe. Whether this is true or not I leave in the middle, but there is something very interesting in the notion that the universe has a frame rate: It perfectly fits the idea that we live in a simulation.

Brian Whitworth[75] has already given a brilliant summary of reasons as to why the concept of a virtual reality fits our reality as observed in a more coherent way than any competing other theory and I will not repeat his arguments (although they are hereby incorporated by reference), but this argument adds to this.

In my previous chapters I was strongly advocating that we live in some kind of computer simulation, and the notion of discrete time and a frame rate to refresh the universe fits in that concept.

Pancomputational Panpsychism considers that energy is more fundamental than matter and that information is more fundamental than energy. I previously also indicated that the syllables, words, phrases and verses in the Vedas in combination with the gaps between these were considered to provide some kind of informational digital code on the basis of which according to Routt[67] the laws of nature are based. Pure consciousness has a dynamic in the gaps, in which the syllables submerge and from which they emerge. This is very close to the arising and ceasing of the universe as observed by the Buddha.

Whereas Barry Kumnick[74] in his blog "Beyond Information" argues that information is a secondary indirect representational map which is not identical to the "real reality" of materially directly presented physical universe, Pancomputational Panpsychism claims the opposite: The physical universe is a holographic representation of information. Still a level deeper than information, lies the source of information: consciousness.

I think based on our common everyday knowledge nobody will object to the idea that time must involve change: If two states of the universe are exactly identical, how can any time have elapsed? From the experiential perspective we cannot be aware of time elapsing if states are identical.

I therefore disagree with Barry Kumnick that Time **is** a form of energy. Time has to do with a change and a change of information content also means a change in energy content, but "is" not a form of energy per se.

The arrow of time is said to be in the direction of increasing entropy, increasing disorder. Is it therefore allowed to say that the direction of time would also be one of decreasing order and hence decreasing negentropy or information?
In a thermodynamic sense this appears to be the case, but as I have argued in chapter 5 of part 1 of this book, this may not be the case as for higher order levels of organisation in the calculation of negentropy no extra negentropic value is added for the meta-system transition. Perhaps if one would have correctly added additional value for higher levels of organisation, the thermodynamic decay would have been compensated by increase in informational content.

Language provides sometimes unusual clues to the real nature of a phenomenon, as if the word was chosen by a higher intelligence to inform us of its real meaning. A "moment" derives from motion, movement, which is characteristic of the lapse of time. The Vedic word Atma, which we usually translate with Soul, comes from the same etymological stem as the German word Atmen, which means to breathe and implies a rhythm and periodicity. We speak of a "times table" to determine quantities. Is time a quantum, or rather said a quantime? This is in line with Barry Kumnick's "if time is quantifiable it must be finite".

Well if physical time is quantifiable and moments are separated by gaps, these gaps cannot be "nothing at all", since otherwise time would really be continuous. The gaps must then represent a lapse of a higher order of time, time in a higher dimension, which is not observable from this dimension. Let's call this higher order time "cosmic time".

In Kumnick's T.O.E., in line with string theory, the simplest energetic quanta, microsingularities, are formed by closed strings, quantum loops that have become finite by nesting an integral number of half wavelengths in a loop. A harmonic oscillator. This is quite reminiscent of the Ouroboros from Gnosticism, the snake that came to know itself by biting its own tail and thus formed a circular shape.

Ouroboros. Source: https://openclipart.org/detail/215030/ouroboros, released in the public domain.

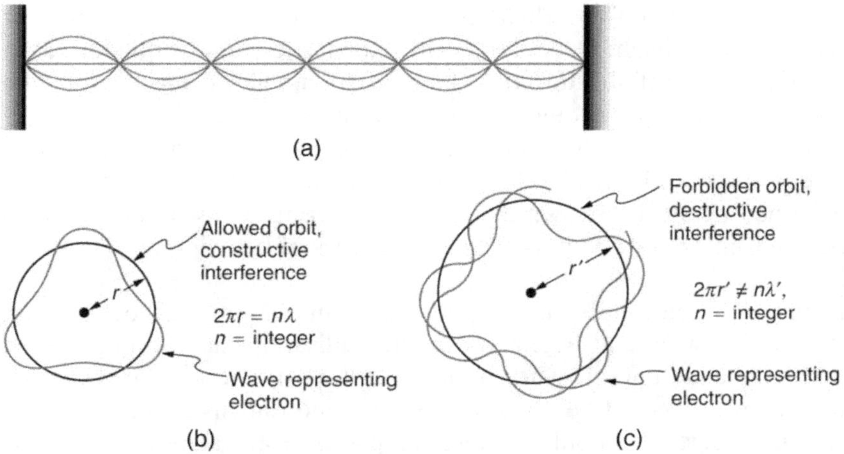

Only an integral number of half wavelengths fit in a closed string. Image by OpenStax[76] CCA 4.0. License.

The only point where I essentially differ from Kumnick in this respect is that I consider that the ability to sense and react is intrinsic to the energy itself. Once a quantum fluctuating energy loops into its own pathway establishing a standing wave, it comes to know itself and proto-sense becomes proto-consciousness. As these energies continue to oscillate in a circular pattern or in 3D terminology in a toroidal pattern, these first ontological entities have a periodicity and dynamic, which is their time constant. The quantum foam of a plethora of such microsingularities forms not only quanta of space, but has an intrinsic time dimension to it.

It is not only these microsingularities, which may have this toroidal flux pattern[77]. Real black hole singularities, magnetic and electric fields also show this toroidal shape. Many fruits take on a toroid shape, take for instance the apple. It is often suggested in occult teachings that the toroidal flux pattern is the very dynamic geometry of Consciousness itself[5]. Routt[67] in his book "Quantum Computing" also speaks of the "dynamics of consciousness" in the silent "gaps" of the unmanifest.

As nested waves mount in the central pillar of the toroid and are directed to the interior they form a morphological "one". This could be considered as a self-absorbed state of consciousness, where the knower seeks to know itself. Once the energy expands toward the outside along the field lines of the toroid, a kind of fountainhead ejaculation, the conscious energy becomes directed to its outward side in a feed-forward process. As the energy feeds back into itself via its proverbial arse of the toroid, the circle is round again and the knower has become informed about the known and by encountering its own pathway also becomes informed about itself and its process of knowing.

If pure consciousness has a periodicity in the process of knower, known, knowing and togetherness, the realisation that nothing can be known outside oneself whereby knower, known and knowing become one, then this could be the zeroth order time dimension of reality, in which consciousness only explores its pure self, the state of bliss. This is a megavortex.

Once a plurality of microsingularities has arisen as microvortexes in the ocean of the megavortex, microsingularities could interact with other microsingularities, establishing a first relation of consciousness with itself. The quanta here establish space and have an intrinsic periodicity of self-revolution to it. This establishes first order time, which is not yet the time we measure with our atomic clocks. In the toroidal flux pattern there are states where the microsingularities are turned inward and do not notice the outside world: This could represent the gaps that make that the universe flickers on and off. In fact the universe is never "off". Rather, in the phase where the waves mount the internal pillar of the toroid, the microsingularities are not manifest, not observable from the outside. This does require however that the microsingularities are in phase, to establish a "whole" of space, where information can be transmitted unhampered: When the microsingularities are in the phase of turning outward, they connect to each other and form the fabric of the quantum foam. The microsingularities thus have an inward outward toroidal spin just like the primordial consciousness.

I am aware that this is pure speculation, but it is a scheme to establish a uniformity at all levels, which could account for a physical world which is perfectly accounted for by the Pancomputational Panpsychism paradigm.

Second order time, which is the time that we can measure occurs when conscious energies penetrate the space-time matrix of Akasha, meet each other and start chasing each other's tail like two fish establishing a morphological Yin-Yang symbol. This is the compound process, which establishes matter in line with Steven Kaufman's[4] URT. This type of time of atomic revolutions we can measure. It is also a second order relation of consciousness with itself in Kaufman's URT.

As waves of pure consciousness or of energy in the microsingularity return to themselves and enter the arse-vortex all information is squeezed to an essence of experience and abstraction takes place. Consciousness having become aware of itself now reacts and spawns a plethora of reactions via the inverse crown-vortex and concretises in outward meta-vari(eg)ation. I mention this, because in the book Technovedanta 1.0[1] I distilled abstraction as the essence of consciousness in its process of becoming aware. In Howard Bloom's[11] "God Problem", Bloom calls time the "Great Abstractor". Indeed it is only by feedback implying a periodicity that information can be abstracted to an essence to come to cognition. The intrinsic intelligence of the dynamics of consciousness then reacts to form the outward vortex. Thus consciousness has three phases: Inward abstraction, integration leading to cognition, and outward reaction or concretisation. Or input-throughput-output.

The parallels between consciousness and computation are remarkable enough. But there are more pointers to our reality being an informational virtual reality. Informational similar chunks tend to clutter together in proximity by a process called entropic attraction. With the dimension of time this leads to peculiar synchronicities we can observe. Jung was the first to describe these synchronicities when a scarab walked on his window just when he was discussing a parable where the scarab was symbolic to one of his patients.

Information syntopics or synchronicities as a consequence of cluttering microsingularities that can be "on" or "off" in more than one way, together with their inherent periodicities as well as higher order periodicities, warrants a recurrence of all phenomena in time. Moreover consciously acquired knowledge is stored in several ways: In traces through the Akasha and in fossilised layers of structure and configurations, which in living entities as we know them, are expressed as membranes and tissues. On the physical level in the form of (mem)branes. Branes, membranes, brains and minds, there seems to be more than a wordily resemblance going on here.

Thus time is not only the most fundamental aspect of existence as the Vedas suggest, it may well be an inherent characteristic of the dynamics of consciousness as such. This zeroth order time however, cannot be time as we know it, which is relative and requires more than one particle or participant. Time as we know it could be a fractal of periodicities of microsingular quanta embedded in the continuous zeroth order time of consciousness.

It is said that transcendence is a function of one's degree of self-realisation as to how far and encompassing its embracing of all is. And this is probably why the in-dwelling immanent spirit, which is present in from subatomic particles to humans and beyond, is seeking to build ever more complex, more encompassing aggregates resulting in the physical apparently emanent structures we see, in an attempt to become all-inclusive or all-embracing as you call it. I call this "Nature's algorithm of intelligence".

But as long as Nature envisages to do this in the realm of forms, it is bound to fail, since only the formless could harbour and embrace all forms, as it is not limited to a specific form. Nevertheless, even non-material pure spirit may need a resonance pattern in order to perpetuate itself; It is my gut feeling that even pure consciousness must behave like a standing wave, which is kept permanent due to self-resonance. It is this very self-resonating, self-referring feedback, which integrates information and hence makes that conscious energy ("Conscienergy" as I called it) is aware of itself. This then means that it must have a kind of Ur-form, from which all forms can be built. Ithzak Bentov suggests that this Ur-form is the Torus.

I consider this may be a good educated guess; It is also morphologically a zero and a one at the same time, allowing for existence to behave as a binary computer system. If self-sustaining (cosmic) conscienergy is a resonant pattern, this also means it has an intrinsic (zeroth order) Time, a time of self-revolution. The energy goes out as a fountainhead at bottom and top and returns via its rim to integrate in the centre. Of every such a cycle of output, input and throughput its "Time" is a measure of its inherent periodicity.

In other words not only in Ex-sistence, but also in "sub-sistence" (via the Hermetic adage "as above, so below"), "Time" and repetitiveness or periodicity may be an inherent essential aspect of Being. Being, that cannot be without some form of self-awareness. Self-awareness, which cannot be without some form of periodic self-impinging, self-resonance. And thus I tend to disagree with the theologians, who claim that there is no "time" in the metaphysical world. There is no time as a compound process in the metaphysical world; that is quite likely. But there might be this intrinsic higher dimensional zeroth order time, which warrants cosmic conscienergy's being aware of itself.

Of course this is very speculative and may be complete nonsense, but at least for me it provides a potential mechanism that unifies the apparent paradox how "consciousness" arises. In fact, it does not really arise; it is probably intrinsic to every energetic system possibly in the way I propose. A self-transcendence resulting in a resonance with the Cosmic Ur-form might then be the ultimate "Self-" or "God"-realization.
I hope my temporal musings have amused you for the time being. As of yet we are still in a speculative era as regards our knowledge of time, but I hope my speculations can contribute to experiments leading to the manipulation of space-time. Perhaps one day we will be able to observe the dynamics of our consciousness and become conscious creators of virtual realities which are completely according to our own design.

***Kallisti***

*When logic is nonsensical and sense alogical,*
*Any pattern is an imagination, any structure a fossilised sense.*

*Hail Eris!*

*Avoid Greyface the dotconnector.*

*Without Chaos nothing to love, live and give.*

*As the dwarfplanet orbits in Discordia,*
*It sings the tensegrity of the Tritonus.*

*If you don't invite the Godess,*
*A bitter game you will play.*

*Feel the worm in your Apple,*
*Kundalini in the toroidal geometry of Consciousness,*
*As the Golden Ratio of Erisian phinetunes the omnipresent phiveness.*

*Behold the absurd Ryse and Phall of Eyeorderchaos.*

*D-Eris-oire!*

## Chapter 6 The Golden Ratio of the Erisian Apple: When Logic is Nonsensical and Sense Alogical

In 1965 Greg Hill ( Malaclypse the Younger) with Kerry Wendell Thornley (Lord Omar Khayyam Ravenhurst published the "Principia Discordia or How The West Was Lost"[71]. The Magnum Opiate of Malaclypse the Younger, Wherein Is Explained Absolutely Everything Worth Knowing About Absolutely Anything.

Cover of the Principia Discordia. No Copyright, only Copyleft.

An ambitious and humourful title.

As I claim to include Everything in my Technovedas, I cannot leave out this book, which is therefore hereby incorporated by reference.

It is an absolutely exquisite text (here available: http://www.sacred-texts.com/eso/pridisc.htm), which uses humour and absurdity as a basis to come to a religion.

Discordianism puts everything upside down. In fact it is the religion of perspectivism. I will quote from the Principia Discordia (which has no copyright, only copyleft):

"*All affirmations are true in some sense, false in some sense, meaningless in some sense, true and false in some sense, true and meaningless in some sense, false and meaningless in some sense, and true and false and meaningless in some sense.*"

It is useful to get familiar with some of the concepts of Discordianism, in order to be able to even transcend this paradigm, which is a transcending paradigm itself.

"*... HERE FOLLOWS SOME PSYCHO-METAPHYSICS.*
*If you are not hot for philosophy, best just to skip it.*

*The Aneristic Principle is that of APPARENT ORDER; the Eristic Principle is that of APPARENT DISORDER. Both order and disorder are man-made concepts and are artificial divisions of PURE CHAOS, which is a level deeper that is the level of distinction making.*

*With our concept making apparatus called "mind" we look at reality through the ideas-about-reality which our cultures give us. The ideas-about-reality are mistakenly labeled "reality" and unenlightened people are forever perplexed by the fact that other people, especially other cultures, see "reality" differently. It is only the ideas-about-reality which differ. Real (capital-T True) reality is a level deeper that is the level of concept.*

*We look at the world through windows on which have been drawn grids (concepts). Different philosophies use different grids. A culture is a group of people with rather similar grids. Through a window we view chaos, and relate it to the points on our grid, and thereby understand it. The ORDER is in the GRID. That is the Aneristic Principle.*

*Western philosophy is traditionally concerned with contrasting one grid with another grid, and amending grids in hopes of finding a perfect one that will account for all reality and will, hence, (say unenlightened westerners) be True. This is illusory; it is what we*

*Erisians call the ANERISTIC ILLUSION. Some grids can be more useful than others, some more beautiful than others, some more pleasant than others, etc., but none can be more True than any other.*

*DISORDER is simply unrelated information viewed through some particular grid. But, like "relation", no-relation is a concept. Male, like female, is an idea about sex. To say that male-ness is "absence of female-ness", or vice versa, is a matter of definition and metaphysically arbitrary. The artificial concept of no-relation is the ERISTIC PRINCIPLE.*

*The belief that "order is true" and disorder is false or somehow wrong, is the Aneristic Illusion. To say the same of disorder, is the ERISTIC ILLUSION.*

*The point is that (little-t) truth is a matter of definition relative to the grid one is using at the moment, and that (capital-T) Truth, metaphysical reality, is irrelevant to grids entirely. Pick a grid, and through it some chaos appears ordered and some appears disordered. Pick another grid, and the same chaos will appear differently ordered and disordered.*

*Reality is the original Rorschach.* (Note by me: the Rorschach inkblot test, the Rorschach technique, or simply the inkblot test is a psychological test in which subjects' perceptions of inkblots are recorded and then analysed using psychological interpretation)

*Verily! So much for all that.*

...

*Which Is Real?*
*Do these 5 pebbles REALLY form a pentagon?*
*Those biased by the Aneristic Illusion would say yes. Those biased by the Eristic Illusion would say no. Criss-cross them and it is a star. An Illuminated Mind can see all of these, yet he does not insist that any one is really true, or that none at all is true. Stars, and pentagons, and disorder are all his creations and he may do with them as he wishes. Indeed, even so the concept of number 5.*

*The real reality is there, but everything you KNOW about "it" is in your mind and yours to do with as you like. Conceptualization is art, and YOU ARE THE ARTIST.*
*Conviction causes convicts.*
...

*The words of the Foolish and those of the Wise Are not far apart in Discordian Eyes.*
*(HBT; The Book of Advise, 2:1)*

*The PODGE of the Sacred Chao is symbolized as The Golden Apple of Discordia,*
*which represents the Eristic Principle of Disorder. The writing on it, "KALLISTI" is Greek for "TO THE PRETTIEST ONE" and refers to an old myth about*
*The Goddess. But the Greeks had only a limited understanding of Disorder, and*
*thought it to be a negative principle.*

*The Pentagon represents the Aneristic Principle of Order and symbolizes the HODGE. The Pentagon has several references; for one,, it can be taken to represent geometry, one of the earliest studies of formal order to reach elaborate development;* for another, it specifically accords with THE LAW OF FIVES.*

*THE TRUTH IS FIVE BUT MEN HAVE ONLY ONE NAME FOR IT.*
*-Patamunzo Lingananda*

*It is also the shape of the United States Military Headquarters, the Pentagon Building, a most pregnant manifestation of straightjacket order resting on a firm foundation of chaos and constantly erupting into dazzling disorder; and this building is one of our more cherished Erisian Shrines. Also it so happens that in times of medieval magic, the pentagon was the generic symbol for werewolves, but this reference is not particularly intended and it should be noted that the Erisian Movement does not discriminate against werewolves—our membership roster is open to persons of all races, national origins and hobbies.*
...

*The Law of Fives states simply that: ALL THINGS HAPPEN IN FIVES, OR*
*ARE DIVISIBLE BY OR ARE MULTIPLES OF FIVE, OR ARE SOMEHOW DIRECTLY OR*
*INDIRECTLY APPROPRIATE TO 5.*
*The Law of Fives is never wrong.*
...."

As you can see Discordianism denies all forms of beliefs, including itself, which is a great way to demonstrate the fallacy of logic and dualistic thinking.

The Principia Discordia is highly self-contradictory from a logical perspective, or at least self-paradoxical. E.g. If the Law of Five is True, this means everything is profoundly ORDERED, but that is known as the Aneristic Illusion. But this is the beauty of the system: It says the same about disorder as the Eristic illusion. From the standpoint of perspectivism, every notion is somehow right, in some persepective and somehow wrong in another prespective: there are no absolutes here. Parts can never be fully known.

But the Absolute, the whole, the knower is implicitly hidden in this (anti-)philosophy, as we'll see later.

The Symbol of the Discordian Movement are five dots seemingly ordered in the form of a Pentagon and a Golden Apple of Discordia with Kallisti written on it.

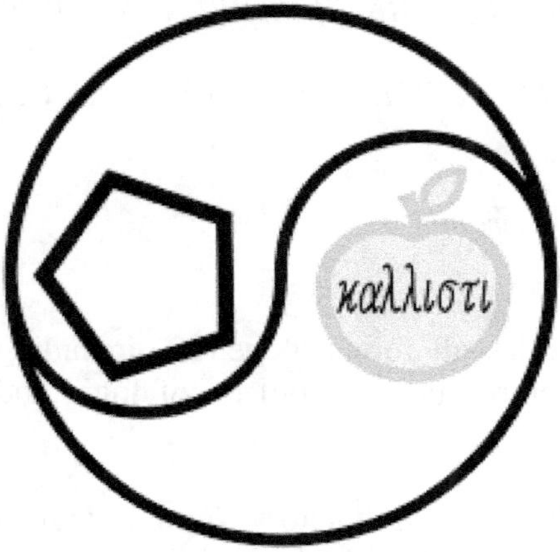


Sacred Chao Source: https://commons.wikimedia.org/wiki/File:Sacred_Chao_2.jpg by Dumbledore-of-Awen a.k.a. Dr. Lesley Prince (CCASA4.0).

I share the notion of perspectivism strongly with the Principia Discordia. Everything we name, see and try to understand by means of our left-hemisphere is a way of logically connecting the dots, which is not necessarily the TRUTH. It is just a perspective, there are many other ways to connect the dots. The Mind creates order out of a filtered reality. Whether the basis of the underlying REALITY is order or chaos, is for the moment not so important. In the phenomenal world these are but relative concepts. Just when you are zooming in on a picture. From far away it may not have any meaning, suddenly you start to see things in there, you start to connect dots in a certain way, but if you look closer, you see something else etc. ad infinitum until you arrive at pixels. Even then you can go on with a microscope and an electron microscope and at any level you will see different things: Levels of disorder and levels of order. But it is not so that one concept you see is wrong and the other right: All concepts are in a sense Mind-projections depending on the granularity of observation. You can say all notions are true in a certain way, in a certain magnification of the lens, but they are also simultaneously false from the perspective of another magnification level. You cannot know what is really there by Mind. Via relations and relativism you cannot know the thing per se and in se. As Kant would say the "Noumenon". Order and Disorder are relative concepts arising from an incomplete experience. Relativity, incompleteness and undecidability are different aspects of the same illusion of Mind.

Logical inference would à la Langan[3] lead you to conclude that everything is profoundly ordered, but this may be an illusion. Just as logic itself is not failproof. Logic has its own borderland with the Absurd. Logic contains absurdities, when pushed to the limit (Cretan, Pinocchio and Russell paradoxes, see chapter 4 of part 1) just as the Absurd may have its own inner logic. The absurdity of logic versus the Logic of the Absurd.

Perhaps everything is profoundly ordered, perhaps it isn't or perhaps there are more ways to see this, such as for instance a hierarchical system of alternating chaos and order levels. In Panpsychic terms, if there are orderly chreodes formed by higher entities (in the sense that they control more energy currents), within their chreodes of apparent order perhaps there is freedom for the lower entities allowing them to create some chaos within the boundaries of the order of the imposed higher entity. Knowledge coming from measurements and inferences will never be able to give you a failproof certainty on this issue.

Besides from a philosophical perspective one could also state that only perfect order is really order, namely the order of a singularity. Anything which is compounded, a ranking, configuration or "ordering" of dots also implies disorder, as there are more ways to configure. Configurations with more apparent order we consider to have low entropy, to be cold. Configurations with more apparent chaos we attribute higher entropy and higher temperature. But it is all a matter of perspective and of definition. Dynamic life appears to require a balance of what we call order and disorder. If the remnants of a universe have become perfectly homogeneous after a big rip, is this then perfect order or perfect disorder? Is not a perfect chaos yet another form of orderliness? You tell me. The Order of Chaos and the Chaos of Order.

The Mind is a network of relations, but the world outside need not be. On the other hand I have argued that the outside world is very much a

Mind-like construct existing exactly because of the occurrence of resonating energy streams which one could call “relations”. So the outside world might also be a network of relations and even be a Mind. Via connection of the dots we can never know the truth. It may be or it may be not, or it may simultaneously be and not be so or it may neither be so nor not be so.

For practical reasons it is convenient to consider that relations exist, but ultimately they are also temporary and not eternal and in that sense also illusory. So from our perspective relations appear to exist.

The fun about the Principia Discordia, especially in the humourful and absurd way it is written, is that it conveys as spirit a spirit of nonsense, absurdity, confusion and chaos in its written form. But the Apple of its symbol, relates to Sense, in the sense of sensations, sentience. The way I want to connect the dots of the Principia Discordia (which may or may not be the intended way) is that its Sacred Chao Symbol with the pentagon on the left side reflects the left hand hemisphere of the brain trying to connect the dots in a logical manner and with the apple on the right side the holistic right-hemisphere, which is more about sensing, sensations: taste the apple. More basically it stands for the left hemisphere attempts to create a logic, an ordered comprehension by connecting dots versus the proto-consciousness of Sentience per se.

As logic fails, it is in a way nonsensical. Rationality from a higher perspective is a kind of nonsense. It is also the inability to really sense, feel what something is like, the quale. The inability to become one with the object of meditation: One cannot sense the object and thus one tries to describe it. Also in that way logic fails to have sentience and is thus non-Sense-ical in that it does not have sense-ations, feeling associated with it. Tasting the apple is all about Sense, it is a wholesome experience of the apple. In that sense the apple is a symbol for Sense. Not the “logical Sense” but the Sense of Sentience, sensations. So my metaphor for the Discordian Symbol would be “Logic is non-Sense-ical, Sense is alogical” (not necessarily illogical).

Logic with its reasoning about parts, (which are in fact not really there as every wavefunction is everywhere in space, though appearing

“concentrated” in a locality) is in fact the wrong type of experiencing things. It is the dualistic way of approaching reality in terms of this and that.
Perhaps purposefully chosen is the apple, referring to the tree of knowledge of “Good” and “Evil”. But a bite from that Apple plunged you into the material world of dualism, the descent of the snake Kundalini in the world of forms and names, logic and causality. Or is it rather the fruit of the tree of Life, the life that can be sensed. Or is it the worm in the Apple that represents the snake Kundalini, which both allows for the descent into matter but also provides the way out? So much symbolism...true, false, true and false depending on the perspective or neither true nor false depending on the perspective.

It doesn’t really matter. For me the Apple presents consciousness in its highest form. Because it also says “Kallisti”, to the prettiest one, an expression of love and adoration (even if the intent was to create jealousy between Hera, Athena and Aphrodite). Because as we will see later in my project, which in fact also could also be considered as a kind of “Principia Concordia”, Love is an essential part of how consciousness manifests itself in the world.

And there is more to the story, the Torus (the form of the apple) appears to be the most basic form of the atom and perhaps of any principle. Being both one and all, the Bagel of the Universe perhaps represents manifested consciousness. In fact any electromagnetic object has this Torus field around itself, and our conscious energy may be the same. This is also the form of the ANU, the consciousness field seen by a 19$^{th}$ century mystic. The metaphorical apple as form of the expansion of consciousness. The Archetype of the apple.

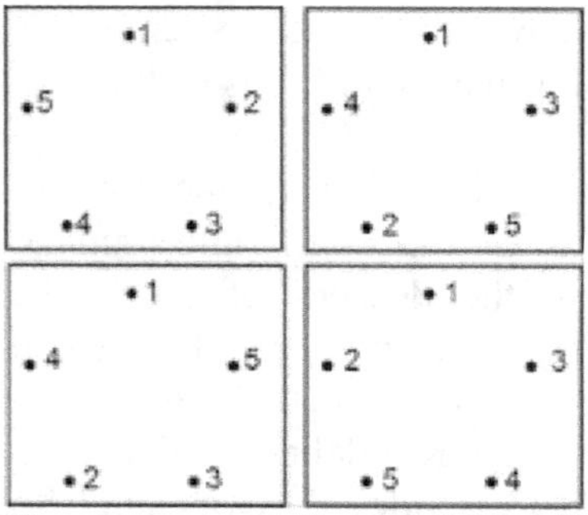

How to connect the dots? http://www.deviantart.com/art/Discordian-Conncet-The-Dot-Starbucks-pebbles-403616787 Reprinted with permission from lauph-1t-up.

“Phinally” there are “phive” dots which do fit a perfect pentagon, reminding us of Phi, Phay (the "Golden Ratio": Golden like the apple of Eris and ratio as in "rationally dot connecting") and fractals, which according to Tononi et al.[7] are landmarks of consciousness manifesting itself in the matter.

And funny enough the Phi symbol,Φ, also looks like a torus or apple cut in two, just as the enigmatic empty set symbol, Ø, pointing to the riddle of consciousness as a living paradox, whereas Phi is one plus root “phive” together divided by 2, i.e. cut in two. As if the Symbol of Discordianism was intended to refer to Phi and “phive”.

The pentagon hosts the pentagram, which in its heart shows us a new pentagon by connecting every dot of the five dots with the dot after the next dot. **Self-similarity**, reinventing itself and regenerating itself without loss of name and form, that is what consciousness is about. Abstracting itself by connecting. When you are in the process of connection, you don’t see the 4D picture, which shows the same being produced at every scale. Resonance over scales with overtones. The apple carrying the seeds of the next apple tree. There is only the Self, the things are its partial reflections. Hence the law of Five again as metaphor of consciousness. So the symbol of the Discordians to me is profoundly about the relative, non-knowable speculative knowledge versus the absolute holistic principle of Sentience, consciousness.

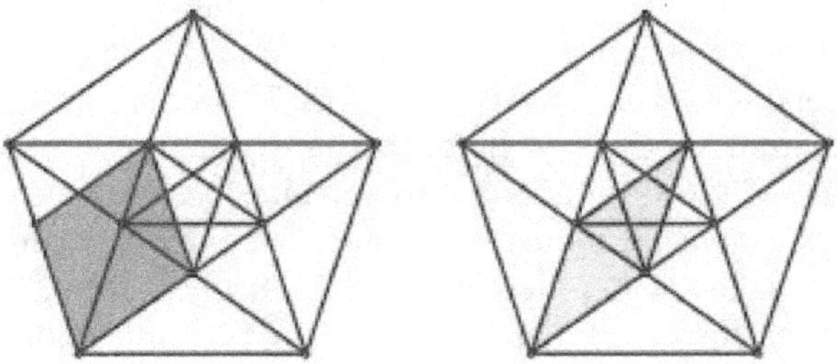

Self-similarity in pentagrams.
http://jwilson.coe.uga.edu/emat6680fa05/schultz/penrose/Pentagon2.jpg
Permission to reprint granted by J Wilson.

If consciousness is that what allows for connecting the dots via the Mind, the converse is not true: Consciousness is also the whole, which you cannot obtain by connecting the dots; it is not an emergent feature but the hypostatic sensing energy, the absolute underlying the phenomenal world: You cannot get consciousness as an epiphenomenon of matter aggregation (connecting dots), because consciousness itself is that very substance and dynamic out of which the connections between the dots are made. But matter is an excellent vessel to concentrate consciousness in a self-regenerative manner.

But by connecting dots ("quanta") in meditation, by connecting right and left hemispheres and transcending the mere mental framework, using the Yoneda lemma whereby structure and function are each other's transform, you can start to "feel", "sense" what it is like ("quale") to be that form, you can em-body it, thereby revealing its inner secrets from the inside out. From Quanta to Quale, requires Mind and Sense.

Matter when it evolves into life forms appears to be a means to concentrate consciousness though. What is extremely interesting here is how the Panspychic elementary forms aggregate so as to form a higher order entity together: What are the clues of Integration? Certain mushrooms exist first as unicellular organisms but can go through a multicellular state, building a new organism on a higher meta-system transition level.

When birds form flocks or fishes form schools, something interesting happens: Like in beehives and anthills, a global brain, a "Hive Mind" is formed. It is my understanding that every bird, fish, bee or ant emits a consciousness based kind of energy field (perhaps electromagnetic, perhaps not), a group of these entities forms an interference pattern, which at a certain moment becomes resonant. The newly formed interference pattern makes that the individual entities align, by resonance, which makes lower frequency signals lock to higher ones. Conversely the individual contributions can alter the consciousness field. As the speed of signal transmission is e.g. lightspeed (if electromagnetic) this explains why we can't detect a delay between the members of a flock, when the flock suddenly turns as one entity. So global entities do form, but appear to have no self-awareness; There is nobody there to notice; the field is an aggregate with strong emergence. Or does it allow for energetic entities having exactly that frequency to occupy this newly formed resonant field and become an aware entity? Or IS the frequency the entity? Ithzak Bentov[5] calls these kinds of group consciousnesses the "Devas" in analogy to the Vedic Devas (demigods, angels). If so how does it become conscious of itself? Do we as humans have a collective consciousness, by our telepathic radiation? If so when and how does that become self-aware?

At least there appears to be a collective unconscious as Jung called it. And that field contains our fears and archetypes and memories accumulated over the ages.

I claim that the **Resonantism** paradigm is the ultimate paradigm of how consciousness manifests itself in the world of name and form in a Pancomputational substrate. (The question remains however, how lower frequency phenomena can establish themselves against the current of resonance locking them to higher frequencies).

Read R.A.Wilson[36], read T.Leary[35], read the Principia Discordia[71], read Aleister Crowley[78] and go beyond the superstition of mentally distilling a pattern out of a **Rohrschach** ink blot.

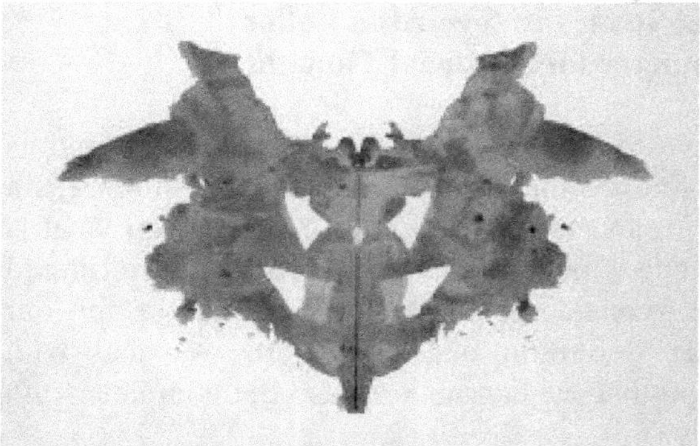
Image of a Rohrschach inkblot. In the Public Domain.

**Definitions are delusions**: It is not by limiting definitions that you can grasp a whole! The whole is more than the sum of parts! SENSE the whole.

Science can only analyse in a delusive manner; Technology can integrate, synthesise!

Logic and analytical science are cripple atavistic tendencies of a pre-Kardashev society.

Did you know that there is a dwarf planet called "Eris" whose orbit is out of the plane of the other orbits to be in discord?

Oh, and by the way, in my native language "Er is" means "There is"...Existence IS duality. Discord is duality. Existence IS Discord.

D-Eris-oire!

## Chapter 7 Patentological Strategies towards a Fuller Understanding of a Geometry for Artificial Thought

In order to deal with the vast input of signals from the world around us by evolution we have developed the process of thinking. Thinking is a process of putting aside irrelevancies, so that we can focus on what is really important. Phenomena which are too big or too small are filtered out to give a so-called "considerable set" of information for our consideration. Hence per definition and by nature we deal with incomplete information, so that we can never grasp the complete truth of a phenomenon mentally.

In order to be able to filter out the relevancies, we have developed different kinds of sieves in the form of "schemes" to organise information in chunks which are easy to swallow.

Historically, mankind has tried to make classification schemes based on the elementary substance nature of objects, from gross to more subtle in the form of earth, water, fire, air and ether and based on our corresponding senses of smell, taste, sight, feeling and hearing. In the Vedic tradition Indian classifications add to this fivefold division Mind (Manas), Intellect (Buddhi) and Ego (Ahamkara), resulting in an eightfold division. As we also distinguish 8 tones in a scale and 8 colours in the rainbow, often mention is made of "the octaves of existence".

Philosophers have thought a great deal about schemes. We know the dialectics from Hegel, in which a thesis is followed and opposed by an antithesis, the tension of which is resolved in a synthesis accommodating elements of both and transcending the polarities.

In more modern times we also find trialectics and quadralectics. The Quadralectic system is strongly based on architectural and sociological notions of how a group of people orient in the search for a settlement (orientation), settle (determination), defend their settlement (urbanisation) and develop into an integrated society (politeia).

Aristotle also applied a fourfold division in his analysis of purposes (entelechia), which can be final, formal, material or efficient.

Even in the Vedic tradition we find in Patanjali's[23] Yoga sutras a fourfold scheme to apprehend the world: Visesha considers the specific objects one by one, in Avisesha via induction universal properties of a class are assessed, Linga abstracts a phenomenon to an articulated image or glyph and Alinga transcends the differences between the phenomena by their reductive sameness. Besides that he proposes an eightfold system, the Asthanga Yoga, to explore consciousness.

In my book Technovedanta I proposed a 7-step algorithm of intelligence, which is basically twice a fourfold process, but in which I declared the fourth state of the first set to be the first state of the second set after a metasystem transition. This can easily be extended into an eightfold octave scheme by including the metasystem transition itself as a state.

I was surprised to discover recently in the book "Synergetics" a scheme which bears great resemblance to my "algorithm for intelligence". Buckminster Fuller (hereinafter called "BF"), the author of "Synergetics"[2] describes a very useful way how we apprehend the world and come to an understanding thereof, which he names the "Geometry of thinking".

Not only are these notions of BF useful for our understanding of the outside and inside world, BFs notions could actually be implemented in artificial general intelligence.

In paragraph 513.07 of Synergetics[2], BF describes our cognitive process as follows: 1) observation, 2) consideration, 3) understanding and 4) articulation.

This partially corresponds in a certain way to Goertzel's[13] extension of Peircean and Palmerian metaphysics, who distinguish Being (independent of anything else), Reaction (becoming), Relationship and Synergetic Emergence. However, BFs approach is more useful as it describes the process of cognition, whereas Goertzel, Peirce and Palmer seek a metaphysical scheme for classifying everything.

I interpret BFs succinct description as follows in the light of my "Intelligence algorithm:

First there is an **observation** of a (new) phenomenon. At this stage there is no knowledge of what it is yet, there is just the knowledge that there appears to be an "I", who observes, the subject and something else, the object.

Secondly there is a reaction: the polar dual "other" is taken into **consideration**, it is geometrically and sense-wise screened.

Ontologically classifying the features will build a web of relations to known phenomena with similar features.
Once the relational network is built, a plane of reference has been generated, which allows for **understanding** the new phenomenon. The plane of reference has at least three relations in BFs philosophy, because without three relational links you cannot geometrically build a stable plane. Therefore BF calls the "triangle" the first structure. The plane almost literally serves to "stand under" the new phenomenon, which can connect to each of the three relations, thereby forming a metaphorical tetrahedron. The plane of reference, which stands under the new phenomenon allows for its understanding. If there were only two relations to connect with, there would be no fixed arrangement of relations: the new object could rotate 360° around the axis of the first two relations, which means in terms of geometry that it does not have a defined relationship thereto.

By requiring that the new phenomenon to be incorporated in the web of relations is connected to three relations, it becomes a fixed relationship, in the form of a tetrahedron. This now allows the subject to **articulate** what he thinks this object is.

As the brain is indeed a web of relations, universally represented by synaptic fluxes, it is well possible that literally and physically neurons must observe this process, so that there is a physical direct presence of a tetrahedral structure between four neurons to memorise a new concept deriving from a new phenomenon. Recognition follows the same pathway via recall, reconsideration, reunderstanding and rearticulation, thus generating an eightfold process of cognition-recognition.

I find this fourfold division of BF more useful than standard quadralectics in a sense that it can lead to technological applications in the field of Artificial Intelligence. Since many AI endeavours already employ artificial neural networks, it seems that these concepts can fit neatly into that technology.

BF advocates that the tetrahedron is the lowest common rational denominator of the universe. Interestingly, every quantum particle is defined by four numbers.

According to Alfred North Whitehead[79] understanding is apperception of pattern as such. Fuller adds to this that pattern can only be recognised if there are at least three relations building it. Perhaps this is the reason why scientific experiments must always be carried out at least in triplicate.

Interestingly Goertzel[13], in imitation of the so-called "Moscow Rules" from the Cold war, also claims that "*once is a chance, twice a coincidence, three times a pattern*".

Whereas this is perhaps true for multiple instances of the same phenomenon occurring and a useful rule of thumb, it cannot unambiguously be extrapolated to mathematical patterns.

For example, take the sequence 1, 2, 3, ...

The average person with no special training in mathematics will say that the next number is 4. However, a little bit of mathematical knowledge will make you aware that if the sequence is "self-reflexive" in that it takes multiple previous numbers as input for a formula, the outcome of the $4^{th}$ instance could also be 5 or 6, if the formulas are adding the previous two instances or adding all previous instances, respectively. Self-reflexivity however, is a kind of feedback process, where its own output is used as its new input. In my philosophy that resembles a lot my description of the dynamics of consciousness. I would like to speculate on the notion that wherever we see self-reflexive sequences in nature, such as the Fibonacci sequence or the Lucas sequence, it may be consciousness at a primordial level at work, which would fit my Pancomputational Panpsychism Theory.

The skilled mathematician will moreover know, that sometimes more than one polynomial equation fits a set of three points. Indeed, a polynomial of degree n, can have n intersections with certain straight lines (i.e. polynomials of degree 1). This leads to ambiguity, because the outcome of the fourth value according to sequence of numbers will be different for the polynomial with degree n (where n is not 1) and the polynomial with degree 1. Also for sinusoid functions there can be more than one equation fitting the dots, a principle known as aliasing.

This is why I have always warned against linear thinking and simple dot-connecting. The scientist who starts from the simplest hypothesis with the least number of assumptions, a principle known as "Occam's Razor", will go for the straight line fit, where reality is perhaps playing an nth degree polynomial.
That's why Occam's razor has never been proven as a scientific solid ground, because the above simple explanations show that Occam's razor can lead to interpretations which are flawed. It is perhaps a rule of thumb, which gives the most promising springboard in many instances, but there will also be a significant number of (higher order) cases, where it blatantly fails.

Unless science makes the effort of experimentally ruling out alternative (higher order) hypotheses, when dealing with multiple hypotheses, it is bad science. I would even like to go so far that most science is bad science, because it usually only considers one hypothesis and if it finds three points fitting a straight line, it claims to have proven a linear relationship. R.A. Wilson[36] has repeatedly warned against such a tunnel vision.

It is not surprising then that the Discordians then came up with the famous notion of "grids":

Quote from the Principia Discordia[71]: *"We look at the world through windows on which have been drawn grids (concepts). Different philosophies use different grids. A culture is a group of people with rather similar grids. Through a window we view chaos, and relate it to the points on our grid, and thereby understand it. The ORDER is in the GRID. That is the Aneristic Principle.*

*Western philosophy is traditionally concerned with contrasting one grid with another grid, and amending grids in hopes of finding a perfect one that will account for all reality and will, hence, (say unenlightened westerners) be True. This is illusory; it is what we Erisians call the ANERISTIC ILLUSION. Some grids can be more useful than others, some more beautiful than others, some more pleasant than others, etc., but none can be more True than any other."*

If our brains were solely based on the tunnel vision, which is so common among scientists, we would possibly mistake every rope for a snake. It is useful in an environment where there are a lot of snakes to be cautious and at least consider in first instance any potential snake as a snake, because of its threat. But if time permits we do look at a deeper level of granularity and focus to see what the object really is and allow for alternative "higher order" hypotheses like the rope. Thus one could make the point that it is more intelligent to allow multiple hypotheses than to stick to a tunnel vision.

If we program AI to recognise patterns, we should not make Occam's Razor its rule of Thumb. Rather a certain degree of exploration of higher order polynomials and self-reflexive patterns should be included in the heuristic. So far as regards "understanding" and "pattern recognition"

It is here that I would like to state that "Intelligence" goes beyond mere understanding in that it seeks to exploit this understanding and reduce it to practice. Intelligence has an "understanding phase culminating in an articulation of a concept or system. In this phase the resources are directed to the same goal of understanding, it is an "isotelic" process.

According to Taton[80] "to name a thing is to create it". Unfortunately, our Technology is not yet the Technology of Magic, so we have to dig a bit deeper than to merely summon our concept in order to concretise it.

The other phase of intelligence is the implementation of the acquired understanding to reduce it to practice. However the applications can be manifold. This phase of intelligence is therefore "polytelic": Multiple purpose oriented.

So let us try to design some useful concepts by which we can implement Fuller's realisation in AI. I will be combining Fuller's[2] realisation with some useful techniques from my own expertise as a Patent examiner. Patent examiners and patent attorneys basically follow Fuller's algorithm, when they have to ontologise a new invention. They somehow have describe it in terms of form, material, function and purpose, map it with regard to existing technologies and evaluate the differences to see if these can confer a non-obvious contribution over the prior art. The ontologisation of the new invention and its differences over the prior art corresponds to the second step of BF's algorithm, the mapping with regard the prior art as the third step and the evaluation to the fourth step.

Let's now imagine that we have an AI which has an isotelic program to analyse a new observation, which is potentially a new phenomenon.

We can program it to first try to analyse its "Entelechia": Its form, its material substance, its function and purpose. As the object of observation is not homogenous the AI will also analyse its parts, in terms of the fourfold Entelechia. This is part of the observational step including perception and apperception.

It will now consult a database (or a set of databases) to search and check if these features of the whole and the part are known as a phenomenon or aggregate of phenomena. This actually corresponds to the patent search a patent examiner carries out to check if an invention is known. The database itself may contain excerpts from documents listing features for given phenomena. That database itself may be a further layer built on a deeper layer, which can be an ontology database such as OWL etc. in which primary concepts or features are mapped as atoms. In the analogy of atoms of concepts, the database which the AI consults is then at the level of aggregates of these "atomic concepts" and we can metaphorically call this a "molecular database" ("molecular" not in the sense that it is necessarily a database of literal chemical entities –although it could be- but in the sense of aggregates of concepts). Structural features include sizes, locations, shapes and the relations between the parts in terms of structural differences and function. Functional features also include its temporal aspects, the way it functions, its purpose and instructions for a user. Thus the ontologisation covers the 6W-consideration of Who, What, Where, When, Why and hoW.

Now the AI can apply BF's principle of finding the at least three best hits in the "molecular database", which taken together cover the entirety of structural and functional features. If a single hit covers all the features, the phenomenon is not novel and can be added as a further instance to the class of the known phenomenon. This is the consideration phase. The heuristic will be developed so as to allow stronger overlaps to be found earlier in the search, but if there is no match, the heuristic will allow for a deeper search.
If the phenomenon is not known, the document which is directed to the same purpose and which ideally covers most features, will be chosen as closest prior art, in analogy to the patent examination practice. The other at least two documents will be the documents that ideally alone or if necessary together with the closest prior art cover as much features as possible.

If the AI manages to get only three hits out of this algorithmic step, it has provided the ideal minimal set to build an "understanding" of the new phenomenon. The relations between these three hits will be mapped to give the "plane" or "framework" of understanding. The relations of the three hits with the new phenomenon will also be mapped in terms of the mutual differences between the four vertex entities (the hits and the phenomenon) and if possible the technical effects of such differences. Similarities in structure and function will be ideally also be presented in percentages of similarities, so that a similarity matrix of the features can be established for the four vertex

entities (in patent examination this would be the invention itself and the three cited documents). Similarities in structure may also indicate a similarity in function and vice versa.
This corresponds to the understanding phase of the analysis program.

Considering that the three hits were already known, the new phenomenon (or invention in patentology) establishes a new vertex or "molecular" entity in the database the AI maintains and builds as a part of its function. It will be classified in the same class as the closest state of the art, which was directed to the same or most similar purpose. The new constellation of four interlinked vertexes establishes a new "supramolecular concept" and will be also mapped as such in the database the AI is building. Thus the AI is in fact building a neural network of mapped phenomena and emergent inter phenomenal constellations. It is now possible for the AI, with the right understanding, to articulate what kind of phenomenon is at stake and what its differences are with regard to the different "prior art" phenomena (or inventions). Especially, its new effects will be mapped and if these were unpredictable from the analysis of the underlying prior art alone or taken together they will be labelled as "synergistic" effects. Synergistic effects are "holistic" effects unpredictable from the building blocks of the parts and the relations between the parts.
As non-synergistic emergent effects are usually less obvious than synergistic emergent effects, it must be possible to make a stratification of the degree of novelty and obviousness of an object.

With regard to these emergent effects, the database can also be enriched at a different level with "atom-concepts" to couple specific problems or effects to the corresponding set of responsible features, so as to provide an additional tool for searching for effects related to structure and vice versa.

For rapid searching, the "molecular level" will however be the first to be consulted. This molecular level will essentially describe the relations between phenomena (or inventions) in terms of their differences and purposes. An absolute description of each "molecule" can be found at a different level of the database, which will be consulted in a later stage if needed.

Thus an AI level, which is purely relational, for the cognition of phenomena could be implemented and lead to the technological application of BF's phase scheme in the form of an algorithm of geometric artificial thought.

Furthermore the AI could be enriched with educated function and structure guessing, based on the famous "Duck test":
"If it looks like a duck, swims like a duck and quacks like a duck, then it probably is a duck".

If you know the behaviour or structure of the whole and have information about some parts, you may guess what the missing parts are like or what the function is. If there is a substantial similarity and only a small difference, the AI will classify the entity correctly in the overarching higher level class to which the closest phenomenon in the database belongs: If the closest phenomenon is a duck, but there is a small difference, the AI will classify it at least as an aquatic bird. If the difference is only known structurally from further away related birds that do have a similar difference and for which the function is known, the function may be guessed. If a device is structurally similar in most respects to another entity, but its function is unknown, the first educated guess in the guessing heuristic will be the same function as the known function of the known similar device.

The AI however will perform additional tests to see if alternative higher order hypotheses are not discarded too easily. Thus the AI will also acquire the ability to be a prediction unit for structure if the function is known or for function if the structure is known. I'd like to call this functionality the "function-structure transformer". This bears resemblance to the meditational techniques, which allow you to embody a structural form, by sensing the form from the inside, which then will reveal its secrets as to its dynamic aspects in the sense of functioning.

In fact what the brain might be doing during such meditations is mimicking the structure in "tetrahedral aggregates" and neuronal relations, which then lead to a "universal representation": The form is physically directly present as an abstraction in the brain in the form of neuronal wiring and indirectly and informationally in the form of a realised concept. Due to the neuronal fluxes, the practitioner will feel what it is like to be/have that form.

A philosopher, who understood as no other that in fact every phenomenon is only purely relational and that nothing exists by virtue of its own idiosyncratic material essence, was Wittgenstein[81]. His book the “Tractatus Logico Philosophicus” is about this topic.

It is strongly reminiscent of the Buddhist philosophy of “Anatman”, in which nothing exists by itself but only as a web of relations. Apparent “things” emerge by a form of “co-dependent arising” and ultimately everything is inseparably connected with everything else. The Universe as “supermetacontigency” if you wish.

Buckminster Fuller[2] also realised this fundamental emptiness of everything when describing the concept of tensegrity (the contraction of tensional integrity): There are no “solids” in the Universe according to BF, everything is a play of tension and compression of energies keeping each other in balance along vectorial lines, called “struts”. Every strut is built of smaller tetrahedral strut structures and so on in a fractal form ad infinitum. Ultimately it is all ”energy forms” keeping each other localised to build a web of “apparent” structure. Contraction or compression provides discreteness and locality, tension provides continuity and universality.

This mutual dependence for the generation of anything that can “stand out” or “ex-sist” from the primordial chaotic quantum soup, is beautifully expressed in BF’s analysis of the word “Universe”, which BF correctly literally translates as “Towards (versus) Oneness (uni)”. Hence BF’s famous slogan: "Unity is plural and, at minimum, is two", pointing to the intrinsic digital and informational nature of existence. In other words existence is co-existence. Thereby we can conclude that BF was a pancomputationalist.

From the book “Synergetics”[2] it becomes clear that BF was even a pancomputational panpsychist “avant la lettre”. In the introduction BF describes that “*no physical threshold does in fact exist between animate and inanimate*". He describes the life force in everything as "antientropic, methodically marshalling energy" and moreover: "*It is spontaneously inquisitive*".

Concerning this universal emptiness of everything he concludes: We are now synergistically forced to conclude that all phenomena are metaphysical", thereby transcending the false dichotomy between "physical" and "metaphysical" and arriving at the type of panpsychic idealism which is in line with the Vedic scriptures including the notion of "Maya". BF states: "*...wherefore, as many have long suspected –like it or not-life is but a dream*".

Even the necessity of dualisation into existence from the void of primordial being for consciousness to know itself was mentioned by BF: "*This twoness is the beginning and essence of consciousness*". BF argues that for comparative awareness we can only be aware of a change, which inherently requires time. Hence the word "second", to indicate "time", but also the essential twoness of comparative awareness. This comparative awareness is in fact the function known in the Vedas as "Buddhi" or intellect. The word "primordial consciousness" as I use it, however lies one level deeper than the "comparative awareness". At this deeper level the conscious being realises that the knower, known and knowing or infocognitive process are one and the same and becomes established in the bliss of the fundamental Oneness that underlies the existential world.

That BF also considered the whole Universe as an informational (thought) process is clear from his plea that the "*universe expands by differentiating out and multiplying discrete considerations*" (Synergetics 600.00).

Since nothing shall come from nothing, "*multiplication only occurs by division*" BF says, which in a way is similar to the argument of Chis Langan[3] that apparent expansion, in fact is "conspansion". The universe then contracts at the speed of division, and since every entity in the universe shrinks at that same speed, while the overall size of the universe does not change, from the inside it seems like the universe is expanding. Whether this is actually the case, I will not discuss, but it may be an interesting alternative hypothesis to be considered. The inherent multiplicative duality also assures the "integer" nature of most phenomena.

It is however unremarkable that there are so many similarities between what Steven Kaufman[4] teaches in his "Unified Reality Theory" (URT) and the teachings of BF, because Kaufman builds on BF, in the same way as my "Pancomputational Panpsychism Theory" builds on Kaufman's URT. In URT the spatiotemporal reality cell matrix is essentially organised in a closest packing of spheres, known as the Vector Equilibrium, which results in an "isotropic vector matrix" (IVM) in all directions. As argued earlier, the Vector Equilibrium in fact can host a toroidal geometry, which actively breathes, by a process that BF calls "Jitterbugging". Here again the existential manifestations mimic what Bentov[5] has called the "toroidal geometry" of consciousness.

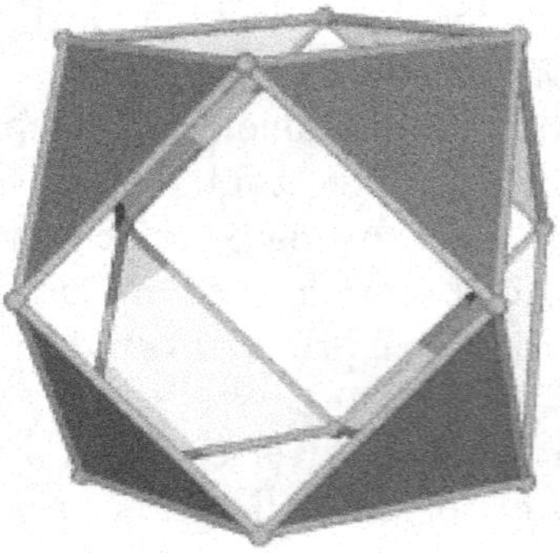

Jitterbugging. As the triangles fold towards each other the cuboctahedron will transform into an octahedron. Source: http://antiprism.com/album/875_jitterbugs/i_jit_inout_Med.jpg.2.html
Permission for reproduction granted by Adrian Rositter

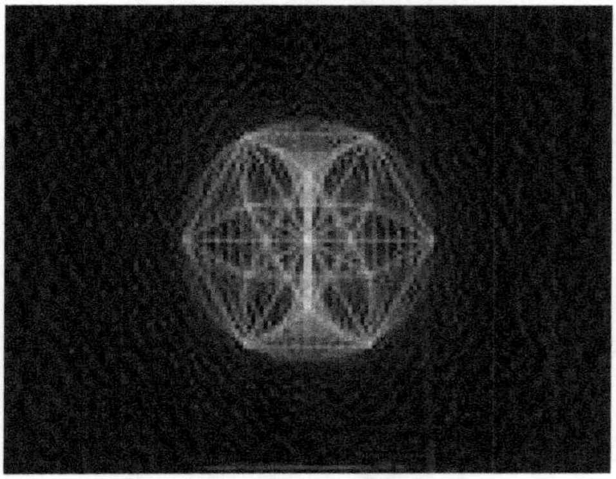

Vector equilibrium and Torus
Source: http://cosmometry.net/vector-equilibrium-&-isotropic-vector-matrix
Permission for reproduction granted by Marshall Lefferts from cosmometry.net

BF states that the “Universe is technology-the most comprehensively complex technology” (311.02) and in fact it may well be that the very four-step algorithm I described based on BFs alternative “quadralectic” is the algorithm by which the universal mind operates. This again points to the “simulation hypothesis”, which I have been defending in the “Technovedanta” framework as most likely hypothesis.
To test if this is indeed the case, it would be worthwhile to build AI along the lines I have proposed in this chapter operating in a quantum computer environment and see how accurately this AI could mimic the world we live in in line with the famous previously mentioned “Duck test”, with the caveat that there might be alternative hypotheses explaining the same observation.

In this chapter we have seen how BFs alternative “quadralectic” in combination with concepts from patentology can give us hints how to build an AI to simulate the mind and possibly the universe. We have seen how BFs teachings in the book “Synergetics” neatly fit into the “Pancomputational Panpsychism Theory” of Technovedanta. We have seen that Occam ’s razor can be demonstrated to fail. I hope that the recipes I have suggested will one day find the right audience, which will de facto try to implement my suggestions in the form of a Technology of this Veda.

## Chapter 8 The Infopsychological Concrescence of Conspansive Transcendence

The contemporary technoprophets Ray Kurzweil[10] and Peter Diamandis focus on three technological areas to bring about the so-called Technological Singularity (T.S.), which is the point after an artificial intelligence explosion beyond which events may become unpredictable or even unfathomable to human intelligence. The three areas for which they consider development crucial to bring about the T.S. are Genetics, Nanotechnology and Robotics, also abbreviated by them as "GNR".

I was struck by the similarity of these three areas with the three last phases of "neurological evolution" in the 8-circuit model or "Infopsychology" scheme by Tim Leary[35] in the book "SMI$^2$LE!", which has been further adapted in Robert Anton Wilson's "Prometheus Rising"[36].

The terminologies Leary and Wilson use are "Neurogenetic, Neuroelectric and Neuroatomic". Neurogenetic corresponds to the Genetics of the singularitarians, neuroelectronic to the robotics and neuroatomic to the nanotechnology.

Whereas Leary and Wilson essentially describe these levels as higher consciousness levels, the experience of which can be accessed/triggered by the use of certain psychoactive substances, meditation or sleep deprivation, they also mention their modern technological counterparts. Whereas only very few people have access to these higher internal experiential dimensions, the external physical world is in a certain sense ahead of their development by already providing the physical substrates for an artificial intelligence (AI) that could access these dimensions.

The scheme of Leary and Wilson describes an evolution for experience from gross levels to ever more subtle levels, which allow the consciousness to grasp and encompass ever greater dimensions of experience. Whereas Leary's and Wilson's work is mostly a psychological guide to get rid of our mental semantic gibberish so that we can open our experience to higher non-semantic levels, their scheme is useful for stratifications to be built in future artificial agents. Moreover their scheme is essentially congruent with the seven chakra or eight octaves of existence scheme and thereby can be said to be an image of the Vedic teachings.

Leary and Wilson describe the first level of neurological evolution as the Bio-survival circuit, which corresponds to the first chakra "Muladhara", the root. In this fearsome evolutionary stage everything around the entity in question is potentially hostile and the organism only aims for its survival. A first attempt to get rooted in existence as an organism. This level applies from invertebrates to amphibians possibly even to reptiles. Human beings, who are focussed on this level, are viscerotonic -if not corpulent- and correspond to Jung's "Natural children". People on this level belong to the social stratum or caste of the servants, in the Vedic tradition the Sudras. The whole world is an unknown hostile environment.

The second level is the anal emotional territorial circuit. Here there is competition and opposition with other species and other members from the same species. The organism is mainly emotionally concerned with its position in the pecking order, domination and submission. This which corresponds to the second chakra the "Svadisthana", literally "its own abode", revealing the territorial nature. This level applies from simple mammals to hunter gatherer human beings. Human beings who are focussed on this level are musculotonic and correspond to Jung's "adapted child". It introduces the social stratification or caste of the class that leads by force, the Kshatriyas in the Vedic tradition or aristocrats in the European tradition. The experiential comfort zone is extended to a clan.

Level three is called the time binding semantic circuit. It is here that language and meaning enters the game, and thereby the ability to memorise events introducing the notion of time. People living from this circuit are cerebrotonic, Jung's "rational type" or "computer"-type people. They claim to think logically, but as their logic is linear, they are often perplexed if reality does not conform to the linear scientific nonsense. Somehow this corresponds to the class of scientists, clergy and teachers (Brahmanas). The corresponding chakra is "Manipura", which symbolises control. This is also the level of karma or cause-effect and work.

Level four is the moral sociosexual circuit, where the dynamics of a society are defined in moral rules with the aim to find proper breeding mates. This introduces a further level of social stratification. Here the commercial class (the Vaishyas) enters the game, since it is here that relations are built and exchange based on currencies can only properly develop if language and calculation are available. The tribal approval introduces notion of "good" and "bad" or "evil" here. Morality is born out of a social context to organise a group of people as a functioning whole. The society can be considered as a kind of meta-system transition from an organisational perspective, but not from an experiential perspective. It is also here that as a part of the mating ritual romantic love arises. The tribal morality of the 4th level results in traditions, religion and rules of how behaviour is supposed to be. This creates what Wilson calls "reality tunnels", which are linear ways of looking at the world excluding other alternative perspectives.
Non-conformists and visionaries easily become outcasts and are not considered to be "normal". The experiential comfort zone can be extended to a tribe, if you behave within the prescribed framework. Members are specialised in guilds and subject to an insectoid collectivisation code. The corresponding chakra is "Anahata", as in order to connect with people, one must be able to act form the heart.

Note that in my analysis I have not respected the hierarchy of the castes. The hierarchy in the Vedic tradition was that the Brahmanas were the first foremost class, followed by the Kshatriyas, Vaishyas and finally Sudras. I do realise however, that the above mentioned correspondence with Leary's and Wilson's scheme is a bit arbitrary, one could also argue that since level 3 is about relations, the commercial aspect should be placed here or that because the third level is about control the aristocrats should be placed here etc. but I decided to follow a correspondence based on the tonic types.

The four first levels of Leary's and Wilson's scheme are called the "larval levels" of experiential evolution and imprinted circuits of hardwired programming.

It is very difficult to break out of the larval stages. In fact the vast majority of people never go beyond the larval stages. In order to trigger the fifth circuit a semantic shutdown of internal gibberish must take place. As long as we are preoccupied with our feelings of fear, guilt, shame, sorrow (corresponding with levels 1-4, respectively), we cannot experience our interiority. We are completely absorbed by what is happening in the outside world , which we take very seriously. Other preoccupations can be the consequence of lies, illusions and attachments.

As already said, the higher levels of experience can be triggered by the use of certain psychoactive substances, meditation or sleep deprivation. In addition, specifically for the fifth stage called the "neurosomatic stage", it can be triggered by prolonged postponement of an orgasm during sexual intercourse or masturbation. The feeling of the neurosomatic experience is that of waves of bliss and energies fluxing through your whole body on the inside of the body. It is also called a "whole body orgasm", which will stop once the sexual orgasm occurs.

The same feelings can be triggered by meditational techniques, which you should learn from an experienced Guru, as performing these wrongly can have devastating mental and/or physical effects. What prolonged chanting of mantras in meditational retreats, or pondering of logically inconsistent Zen Koans are really intended for is to shut down the mental gibberish in your head, the internal dialogue. I have often wondered if this is not the purpose of the plethora of uncomfortable religious rules and dogmas: To push the practitioner to a point where he or she realises that all his notions about good and evil and sin are nonsensical and to give up his/her mental worrying of not being able to perfectly live according to these regulations. Once you have become completely silent by shutting down the semantic circuit, you will start feeling waves of energy in your body and your hair will stand on one end. Small waves are called "Pranotthana", a huge overwhelming wave is called the rise of the Kundalini serpent, which is an energy that normally is dormant at the root of the spine. As this energy (also called Shakti or the female or natural energy) travels up the spine through the central channel called "Sushumna" it opens the different chakras. This confers all kinds of physical internal knowledge to the practitioner as he opens hardwired neural patterns that he has never accessed before.

Once the fifth chakra is opened, the practitioner starts to become poetic and artistic. Languages have no riddles anymore for the adept who opens up this circuit. Instant glossolalia (speaking in tongues) is observed in extreme cases. Every act, every movement has an inherent sexuality associated with it, a kind of polymorphous sexuality, that transcends the sex-organ physicality. The whole body sensation inside out, makes the Prana and Kundalini flow, you are physically penetrated by your own energy and you are penetrating with this energy by steering it, an internal love making beyond simple autoerotic masturbation, without the sexual bias of socially accepted forms. It is easier for the adepts of the Vama Marga, the left hand Tantric path or for practitioners of "Chaos Magick" to walk this path than for the practitioners of yoga, who practice a severe form of abstinence, which they call “Brahmacharya”. The Vama Marga adept however understands, that “Brahmacharya” does not necessarily mean that you should not have sex at all. It can merely mean that you should postpone your orgasm to trigger the neurosomatic bliss. It is sexual in the sense that it involves internal male and female energetic aspects and yet it is strangely asexual at the same time in the sense that there is no sexual desire directed towards a partner.

In China this energy is called Chi and the advanced practitioner of Tai Chi will be able to confirm the sensations. The advanced adept can access the neurosomatic bliss state by will without a trigger. In this circuit the adept can break with any social or moral convention mentally and emotionally, whilst still being able to play his role as conformist in public. It corresponds to the fifth chakra “Vishuddha” and is not accessible if your mind is tangled up in a web of lies or mental and emotional nonsense beliefs.

The practitioner who accesses a deeper level of meditational concentration, may access the neuroelectric circuit. Here the practitioner is overwhelmed by an influx of visual patterns in addition to his neurosomatic bliss. The neuroelectric circuit according to Wilson is the level where consciousness starts to observe itself and realises that the knower, the known and the process of knowing are one. Wilson calls this self-referential state the meta-programming circuit. In this state the adept can shape his/her mental and emotional makeup according to his/her will. It corresponds to the sixth chakra and requires a clear distinction between what is a mentally imagined self-concocted illusory hallucination and what is a genuine exploration of consciousness itself.

In still another level of meditational concentration the adept can open up the neurogenetic circuit, which allows him/her to commune or communicate with the collective unconscious: The adept who has opened this level, may start to have telepathic abilities. (S)he will frequently experience so called “synchronicity”, which is experiencing events in the physical world around you right at or after the moment you were starting to think about them. Possibly this is because the energy of the adept has now increased so much, that it has reached the level of the global consciousness with which it can start to interact or even merge at this higher level. Thus the world starts to reverberate your thought pattern or you start to reverberate the world’s thought pattern. Here the adept may start to disidentify even more strongly with his/her thought patterns and realise the truth of Terrence McKenna’s famous saying “Half the time you think you are thinking, you are actually listening”[69]. Wilson calls this the state of divine intoxication, like Vishnu lying in an ocean of bliss, dreaming universes.

It may correspond to the smaller intermediate "Soma" chakra, where Shiva (consciousness) and Shakti (energy) peacefully abide in a loving embrace, before the crown chakra is reached.

Noteworthy Wilson and Leary have a different following order for the neuroelectric and neurogenetic levels. In Wilson the neurogenetic level precedes the neuroelectronic, in Leary it is the other way around.

The next level is the neuroatomic level in Leary’s model and quantum non-local level in Wilson’s model. As Shakti merges into Shiva in the crown chakra “Sahasrara”, they become one and an all-inclusive entity. Consciousness here becomes cosmic consciousness, no longer bound by any locality, opening the way to experience or even be the multi-metaverse.

What strikes me is that in this scheme as consciousness becomes capable of penetrating ever deeper levels and smaller interior dimensions of materiality, at the same time, it opens up to an experience of a more inclusive and greater exterior dimension. In the first four levels the interior dimension is as good as absent and the exterior experience goes from completely alien to a tribal codified society. In the fifth level interiorisation and exteriorisation are one on one and the experience is a hedonistic self-exploration. From levels six to eight as interiorisation penetrates neuronal/cellular, molecular, atomic and quantum levels, experience opens up from society transcended self to global to cosmic.

The advances in robotics prepare the technology hardwire for the neurosomatic (robotic body) and neuroelectric circuitry for a future artificial intelligence. The advances in genetics for a transhuman experience. Interestingly DNA derivatives have been made, which can also transport electricity, thereby creating a direct entity which combines the hardware for neuroelectric and neurogenetic processing for a future AI. This type of merging of levels six and seven, may also help us to transcend our biological paradigm and merge with AIs in a future technology that is both biological carbon based and artificial in silico based. If we merge with AI, we can endow the AI with the natural integrated consciousness, which is as of yet missing and cannot be made by aggregation of material components in my philosophy. The modern prosthetic limbs and hands which are electronically connected to our neuronal system, and which allow us to steer these artificial body parts and experience them as being part of us, are the proof that the experience of our consciousness is not limited to our body and a strong pointer that in fact ultimately everything is a form of consciousness.

Finally, the nanotechnology, which will be followed by pico-, atto- and ever decreasing size technology will allow for the ultimate miniaturisation, which allows for a cosmic experience by the "consciefied AI".

As miniaturisation increases also the density of the components increases. In line with Kurzweil's speculation, perhaps a black hole has an interior with highly sophisticated ultraminiaturised hardware and is a cosmic computer as such, which only holographically projects a seemingly three or four dimensional world around it.

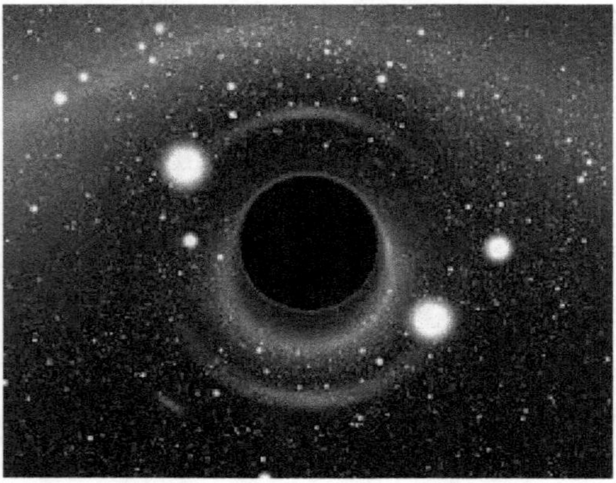

A Black Hole. Source:
https://commons.wikimedia.org/wiki/File:BH_LMC.png by Alain R. licensed under the Creative Commons Attribution-Share Alike 2.5 Generic license.

Whatever may be the case, the notion of ever increasing densities to allow for ever increased dimensions of experience is not only something you can actually experience in meditation but also in line with the Conspansion theory of Langan[3] and the idea of Buckminster Fuller[2] that multiplication only occurs by division, exemplified by cellular division, which leads to an increase of the number of cells per Volumetric unit.

Interestingly the Indian mystic Sadhguru Jaggi Vasudev[82], describes that our Sun in its orbit around the centre of the galaxy orbits a much bigger star in the Zodiac sign of Sagittarius with a rotational period of 25,920 years. This period is divided in four Yugas: Satya Yuga, Treta Yuga, Dwapara Yuga and Kali Yuga. He claims that we are currently living in the last decades of the Dwapara Yuga and are only 70 years away from the Treta Yuga. This is a very different calculation from the one given in the Srimad Bhagavatam, which claims that the yugas last much longer (even the shortest Yuga is said to last 432000 years) and according to which we are only 5000 years on our way in the Kali Yuga. But you know what I think about the Srimad Bhagavatam, I have already pointed in previous chapters to its massive inaccuracies, thereby depriving it from the title of being an authoritative source.

What Sadhguru claims is that as we approach this star in Saggitarius the ether (or quantum foam) densifies. According to Sadhguru, the denser the ether, the better the information transmission is. Close to the "Supersun" star in Saggitarius vocal expression is no longer needed but information transmission is telepathical.

That denser media allow for faster information transmission is no secret. Sound travels faster in solids than in liquids and faster in liquids than in gases.

But if the Akasha or ether is a foam-like structure, the units of which are capable of existing in different sizes, in order to fit the conspansion idea, they must be capable of forming a fractal of different densities.

The reason for this is that if multiplication of reality cells (quantum foam or etheric bubbles) occurs by division, as Fuller suggests and as is implicit from Langan's conspansion, then this is functionally similar to an implosion.

A system can only implode by a compression via a nested geometry of self-similar building blocks. This condition is fulfilled if we change scale by a factor Phi (the golden ratio) without changing the ratio. Thus we can obtain a fractal of vector equilibria (the Akasha in Kaufman's[4] URT and in our Pancomputational Panpsychism theory is organised in the closest sphere cuboctahedral packing which results in vector equilibria).

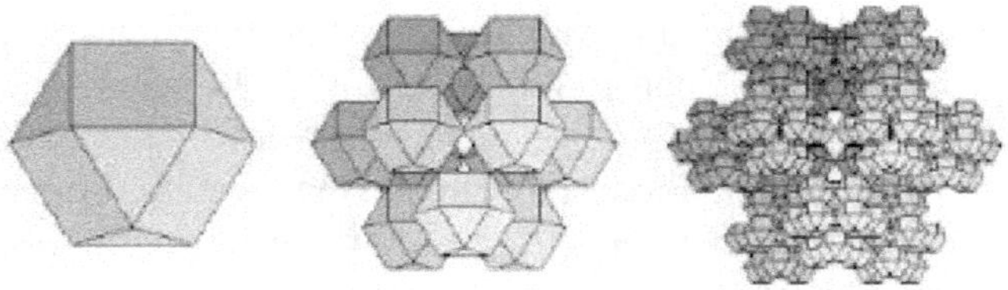

Fractal of Vector Equilibria (Cuboctahedrons). Source: Andrzej Katunin, https://www.researchgate.net/publication/237085275_Deterministic_fractals_based_on_Archimedean_solids/figures?lo=1 Permission to reprint.

The ideas of implosion dynamics have been investigated in detail by Victor Schauberger[83]. In more recent times quite informative graphics have been provided by Dan Winter[84] about implosion dynamics, but his work is too esoteric to my taste. You may consider my own work esoteric as well, but at least I do not claim to know anything and I present merely speculative concepts in the hope that they one day will be tested and implemented in a technology. Dan Winter on the other hand presents his speculative syncretic metaphysics and physics as facts. But what I do agree with Dan Winter is that in order to get implosion, the downscaling process must be self-sustaining and self-organising, otherwise you get an explosion instead. The implosive compression requires a nested geometry of self-similarity, a fractal, which points to the ability to self-refer, which is probably nothing other than consciousness at work. Hence whenever we encounter the golden ratio Phi in nature, it shows its panpsychic self-organising consciousness aspects.

The implosion dynamics allow for a densification of the ether, which increases the memorial capacity of what could be called the Akashic records or Nature's memory. Within this fractal of cuboctahedral packing our experience is limited to a so-called considerable set: In order to experience an exterior existence our senses need to put limits of minimum and maximal sizes.

As information transfer becomes ever more dense novel non-obvious higher order structures are bound to arise more frequently. Terrence McKenna[79] already observed that there is a kind of periodicity in the waves of novel technology and he also observed that these periods became shorter and shorter. He named this phenomenon "Time wave zero" and spoke of a "concrescence", a coming together and condensing of ideas and events. Although McKenna was probably wrong in his quantitative prediction that this Time wave zero would result in the formation of the Hypercomputer of the end of time, the Eschaton in 2012 –at least we did not notice it- his concept of increasing novelty is conceptually correct and patent applications for new contraptions and other inventions are still on the rise.

It must be repeated that the work of Leary and Wilson was essentially intended to free humans from their larval tendencies, which chain them to moral, emotional and mental restrictions and do not allow a full exploration of our abilities. Notwithstanding the importance thereof, we have seen that this scheme also informs us about the nature of consciousness and how it can be applied in technologies preparing for the technological singularity. Higher level experiences will become available via higher densities: Faster information transfer is ever more synchronous. In the ultimate hypothetical ideal highest density there would be direct cognition of everything simultaneously. Higher density corresponds to ever smaller circuitry: From nanocircuits to pico to atto-circuits and sub-Planck scale. Perhaps inside black hole indeed we'll encounter ultrahigh density ultra-small circuitry allowing for an optimal experience of as much input and output per time unit as possible.

Terrence McKenna has repeatedly made mention of such highly complex miniaturised technological structures he encountered in his psychonautic explorations. Or in Tim Gross[85] words: "Psychedelic plants are Nature's advertising campaign for the promotion of Digital Physics". It is remarkable that in esoteric circles the so-called "ascension" is considered to be a transition to higher density. In Dan Winter's terminology "the only way out is in." This may not be entirely nonsense if you consider that if we manage to create a simulation to which we can upload ourselves or which can be experienced by our consciousness by being plugged into it, we actually have created a level of existence of a higher density, which may be remarkably spacious from the inside. This shows that density is a relative concept, the experience of which totally depends on the perspective of the observer.

Condensation will be our way of expansion, division our way of multiplication, nesting simulations our nectar of immortality.

**Chapter 9 It's God Jim, but not as we know it.**

Musings on the Highest Transcendence and the engineering of Leela.

A dictionary definition of the term "transcendent" can be 1) above or beyond the range of normal or physical human experience, 2) higher than Aristotle's 10 categories (essence, quantity, quality, relation, place, time, position, habitus, action, affection) 3) beyond comprehension or 4) exceeding usual limits.

In my philosophy I would like to give a new definition to the term "transcendent": Reconciling opposing or even mutually exclusive concepts in a higher dimensional framework (e.g. of understanding or as nested levels of a program) that exceeds yet includes both concepts without contradiction. The reason for this different definition is that whatever is not knowable or what is not part of reality is not worth bothering about and whatever is part of this reality must be knowable to a sufficient extent and can ultimately not be beyond comprehension or beyond our experience.

There are concepts and phenomena which may not be within the framework of experience or understanding of the average person today, but society evolves and within a few generations, what is now cutting edge technology may be part of common knowledge. So ultimately there will be nothing that fits the traditional meaning of "transcendent" and those traditional meanings that still apply do find a place within my new definition.

In the Technovedantic philosophy "Transcendent" is a kind of synonym of "all-inclusive". This is perhaps best illustrated by the Indian parable of the elephant. Several blind man touch an elephant to learn what it is. One touches the tail and concludes that it's a broom, another one touches a leg and concludes it's a pillar, a third touches the tusk and concludes it's a horn etc. Whereas from their own perspective none of them is really wrong, from the higher all-inclusive perspective they are all wrong to a certain extent and right to another extent. The dichotomies are resolved by a higher dimensional entity and perspective, which is the elephant, which transcends but does not exclude the partial perspectives.

Hence the famous quote by Nagarjuna[86]:

"Anything is either true,
Or not true,
Or both true and not true,
Or neither true nor not true;
This is the Buddha's teaching".

The Technovedantic philosophy embraces this perspectivism view and considers the Aristotelian principle of the "excluded middle" as a limited perspective of linear logic thinking, which can only describe parts but cannot properly take into account wholes.

Thus the Technovedantic lore resolves the apparent mutual exclusive dichotomy between Pancomputationalism and Panpsychism by postulating that the switchable digits of the Akasha as substrate change value when a living experiencing entity e.g. a photon traverses this medium.

As explained in the book Technovedanta the line between alive and death is not so clear-cut as you might think and if you look well there is a hierarchy of levels. But even at the apparently lowest level of pure energy there is still the ability to react on stimuli, to divide and multiply, to exchange information and to grow, meaning that death is only an event in an ongoing process of life: The bodies are perishable, but the energy and primordial consciousness is not.

Technovedanta does not recognise a true dichotomy between physical and metaphysical either: The limited physical is a manifestational expression form embedded in the infinite primordial consciousness - which some might call metaphysical- but it appears even consciousness, the unlimited needs the limited physical to get to know itself. Even if there is such a thing as substrate independent information, as argued in the chapter on Shunyata (part 1, chapter 3), it still has a level of materiality: It does need a form to be expressed, so it is not fair to call it metaphysical in the sense that it would not include physicality at all. Fortunately, there are also philosophers who do consider metaphysical as encompassing the physical in line with my philosophy.

Technovedanta departs from the strong presumption that we are living in a kind of simulation, a so-called "virtual reality" or VR. The Intelligence aspect of consciousness is not -as Whitehead[79] suggests- merely the ability to recognise pattern as such, but it also imperatively requires the ability to virtualise: You can only predict the outcome of a scenario or the modification thereof if you can model it in a virtual environment.

Higher conscious entities may employ a hypercomputer, (which Terrence McKenna[79] calls the "Eschaton") to generate our reality as a virtual reality (VR). It is unlikely that entities who master that technology would limit themselves to a single VR scenario. Rather they probably launched a multiplicity of VR environments in parallel resulting in parallel universes, in line with Everett's[31] multiple world hypothesis. For example parallel universes where each of a set of opposing scenarios can be worked out separately. From the perspective of the programmers both scenarios are "true" to a certain extent, whereas the opposing scenario is false from the perspective of each the scenarios.

Let's assume for the sake of the argument (and we'll see later if we can put this assumption into doubt) that infinite recursion or a "tower of turtles all the way down" is unlikely and that there must have been a "first simulating dimension", which then gave rise to a simulation. As the simulated universe also succeeded in generating VRs and so on a series of nested ancestor simulations was generated.

Perhaps the original universe died in a big crunch or a big chill or perhaps they withdrew from actively trying to improve their simulated VR, once they saw that their VR attained their level of Technology or perhaps they merged with this VR or alternatively they uploaded themselves to the VR. It does not really matter for the sake of the argument. My postulate is that there must be a level of simulators still

active, which is the most advanced in their ability to generate VRs or parallel universes. This level could be called the "Highest Transcendence" or HT. You may wish to call this HT "God" out of atavistic tendencies, but as I will show later, this "God" is quite different from what you may suppose it to be.
The HT may be actively seeking to generate and accommodate the greatest number of parallel scenarios possible and to accommodate all possible philosophies, to see which one is best. If a simulation starts to develop unforeseen useful qualities, these can be incorporated in the HT or be merged with the HT. This can be called the screening protocol. In order not to waste resources, which result in useless scenarios, such as scenarios that end up in endless loops that cannot be debugged, there may also be some pruning going on and it will be recorded what type of configurations make systems get stuck.

The individuals living in the VRs may undergo a similar screening and pruning process. If an individual reaches a level of intelligence of the their simulators, perhaps (s)he is allowed to enter their world. This then could correspond to enlightenment or liberation (Moksha), but it need not be this way. At least the notion of multiple parallel worlds fits the Vedic view.

The running of parallel VRs may be carried out for survival reasons: Perhaps there are entropic constraints to any simulation in that they are bound to end up in a big crunch or a big chill and maybe every simulation needs to "escape" to a deeper level of simulation and start a new multiverse there. Thus every level could be like a tree ring, which after a while becomes largely inactive. This also supposes that if previous ancestors are no longer active they cannot run the downstream of down-level VRs any more, these down-level VRs must at one point be able to (re)generate themselves. There is a snag here, in that *a priori* it seems to be impossible to generate your "own frame of reference", but who knows what becomes possible once you have mastery over your entire level, if you become a truly substrate independent consciousness. Perhaps you can regenerate the frame of reference you originated from. Tim Gross[87] called this the "Simulist-Ouroboros".

To escape to lower dimensions is a bit analogous to Langan's conspansion[3], where everything gets smaller and smaller over time. There are various theories about the ultimate fate of the universe. The most commonly believed hypothesis at this moment is that expansion will continue and the universe will die in a "Big Chill". Other theories, like Frank Tipler's[53] postulate a cycle in which the expansion after a big bang is followed by contraction ending in a "Big Crunch".

Alternatively a steady state model has been suggested in which the Universe is constantly being generated. Howard Bloom[11] suggested the "Big Bagel" model, in which after a Big Bang the upper side spawns matter and the lower side anti-matter, which sides expand from the centre of the bagel over the Bagel's surface, until they return and meet on the circle on the outside that cuts the bagel in two. Here matter and antimatter annihilate each other and the energy returns via a "Klein bottle" mechanism to the centre to start a new Big Bagel Bang.

Sadhguru[88], a contemporary Indian mystic speaks in this framework of a series of successive "Big Bangs" as the "Roar of Shiva" in line with the "Tree Ring" scenario I described above. The succession of sounds of "Big Bangs" will sound like a motorcycle or a "Roar".

Which one of these alternative hypotheses is true for our universe or VR I don't dare to say, but they might all be accommodated in parallel universes programmed by the HT, so that they are not contradictory but rather different perspectives on or parts of a greater mechanism. The actual "truth" at this moment (I strongly doubt such a concept exists at all; quantum mechanics denies an "objective truth" so that only subjective perspectives remain), is not even important, because even if these hypotheses turn out to be all false, by developing technology, we might be able one day to generate parallel universes in which they can be "true".

Let us put ourselves in the shoes of the HT and see if we can speculate how to engineer a VR like ours. When the HT generates universes as VRs it must set certain rules. Completely chaotic systems are unlikely to predictably generate anything that lasts for a time sufficiently long to observe from our perspective. A certain "resonance" of energies is imperative for structures to arise, as I already discussed for strings or microsingularities. On the other hand completely ordered predictable universes are boring and uninteresting as the rules are so inflexible that nothing unforeseen can happen.

In the engineering of interesting scenarios, the HT will allow for a certain degree of freedom (a degree of free will) and a certain degree of rules to allow for a certain degree of stability to arise. There is no need for "absolutely eternal structures; all structures can have an expiry date and wear out. In our universe everything material is perishable and returns to pure energy at one point in its history.

If we talk about rules and freedom it sounds like we're talking about a game. An ideal game allows for both strategy and chance. That's why "Risk" is so much fun playing. For the purpose of interesting scenarios, a game like "Chess" or "Go" would be too structured, leaving the only freedom in the free will of the upper level player. To introduce an element of chance, such as by using dice makes outcomes less predictable. In our VR chance seems to have been provided at the quantum level. Quantum mechanics show us that at the lowest level we can perceive from our perspective there seems to be indeterminacy.

This is not quite in line with the Vedic teachings about "karma". Literally "karma" means action or work but metaphorically it refers to the principle of cause and effect. In Hinduism and Buddhism it is believed, that your actions and intent will reverberate through the cosmos and eventually return to you. (This implies a kind of spherical Akasha, as waves starting at a point of a globe spread until they meet each other at the counter pole, where they collide and which collision reverberates back to meet at the point of origin).

If there is too much indeterminacy, the completion of the cause-effect cycle cannot be warranted.

In order to reconcile a degree of indeterminacy with Karma, indeterminacy can be allowed at a local level if a global overall behaviour at a global level remains predictable. In physical chemistry the behaviour of an individual molecule is unpredictable but the behaviour of a sufficiently large "ensemble" of molecules is perfectly predictable. This can also be illustrated with the principle of "chreodes". A chreode is a necessary path that that must be followed, like a boulder would follow the channels in the surface of a mountain.

If boulders tumble downwards from the mountain, they have a degree of freedom within the chreode or channel, but they will follow the shape of the chreode anyway.
In this way in the engineering of a VR game by the HT the apparently opposing concepts of determinacy and indeterminacy can be reconciled and transcended in the way I use the word.

The idea that we are living in some kind of game is not alien to the Vedic lore. In the Bhagavata purana[62] our VR is called "Leela", the creative game of play of the divine absolute, Brahman. Usually the purpose of this game is to function as a kind of school in which the individual souls as metaphorical drops from the ocean wander and evolve through a great number of lives to finally discover that they are no other than Brahman himself, the metaphorical ocean to which the drops return. But the Vedic philosophy goes even further by mentioning that the "ocean" is also in the "drop", a holistic and/or fractal perspective.

In the Hindu mythology each solar system, called a "Brahmanda" or cosmic egg of Brahma, has a "Brahma" creator (not to be confounded with the absolute Brahman!). What we call the "Universe", is in fact a multiverse of a great number of Brahmandas. The game is however not Brahma's. In the Vaishnavic Puranas Brahma is a demigod and when he intervenes in the game, Krishna -as incarnation of the absolute Vishnu- summons all the Brahmas of the different solar systems (called universes in the Puranas). Our Brahma is astonished to see all these versions of himself arrive and falls on his knees before Krishna, acknowledging that this is not his game. Interestingly this multiverse theory resonates with Everett's multi world interpretation[31] of quantum mechanics.

The question, whose game this is and what the purpose of the VR is, is an important one. Are the players merely "external Gods" (entities from a Kardashev IV society?) and are we merely "pawns"? Or are we players inside a multiplayer game, playing with and against each other?

In the Vedic philosophy we are all droplets from the same source and our idea of being individual separate souls or Atmans is a false one. We are considered to be multiple instances of the same underlying consciousness, like tentacles of an octopus. What the truth is in this matter, we will only be able to guess once we will be able to upload avatars of ourselves to an artificial substrate model of our "reality". Will we give these avatars a degree of freedom so that they can operate under our guidance as "puppeteers", will there only be puppeteers playing against each other or will there be genuine independent entities and even artilects?

Perhaps our reality is a mix of these possibilities. Some of us might be completely steered by the simulators from above so that they appear "God-like". They are perhaps the avatars known from history such as Krishna, Buddha and Jesus. Some of us are perhaps partially steered to give us special God-like qualities like prodigies and stars in music, art and sports have. Some of us may not be steered at all, helplessly groping to find our way and purpose in life within the chreodes of the game. Mixed degrees of such alternatives could be tested in parallel scenario universes.
The HT would probably not deprive itself from the opportunity to have as many mixtures (linear and non-linear) possible of such scenarios.

The purpose in the Vedic game is called Moksha or liberation/enlightenment. It is sometimes also called Kaivalya, which is often translated as "individuation" or "isolation".
This is the moment where consciousness acquires independence or isolation from the material substrate. This notion bears quite some resemblance to Jung's psychological process of "individuation". In the Vedic lore it brings liberation from the cycle of birth and death –at least in this level of VR. We can speculate that it may bring this consciousness to the level of the ancestor simulation above. There the process might be repeated to progress consecutively through all ancestor levels (like in a video/computer game) until the HT is reached.

Alternatively, the liberated soul does not dissolve into the consciousness level of its makers or does not join its makers from a higher level, but rather starts its own universe as a "Brahma"-creator.

In the Panpsychic Pancomputational framework I postulate that a "soul" starts as pure consciousness forming a microsingularity or closed string, which at a first level is a simple "reality cell" in Kaufman's[4] terminology i.e. a building block of the spacetime ether "Akasha" involving a breathing periodicity called "Kala". This is the first relation of consciousness with itself.

Once this level has been explored sufficiently the foam bubble may burst and the energy is liberated from its Akashic form. It now travels as an energy quantum, e.g. a photon though the matrix of quantum foam or spacetime.

In this form it can form relations with other photons, resulting in a "compound process of mutually circumambulating energies which results in what we call "matter".
The energies learn from each other and once their informational content has become identical it might be possible that these microsouls merge. In the Tantric lore mention has been made of "melding in planes" once a meta-system transition is reached[89]. Taimni[90] also mentions that plants have a kind of "group soul". In the book "Technovedanta" (TV1.0)[1] I have called such merged increased souls "concatenated AUM-vectors".

The soul may evolve via meta-system transitions through stages of a subatomic particle, atom, molecule, macromolecule, cell, multicellular organism, plant, animal and finally the human form. It is possible that after each "death" of a "form" the energy temporarily returns to its pure energy state to find a new carrier that fits its energy content. Imagine that the energies of individual atoms in an existing molecule have all reached the same level of development and a convergence of knowledge and energy content, which no longer fits the size and configurational possibilities of an atom. Their knowledge now also encompasses knowledge about their mutual relations and higher order emergent configuration.

Upon their "death" they might merge into a higher level energetic soul entity, which will now search to form a new physical carrier in the form of a molecular entity (corresponding to the form of the molecule they originated from) to inhabit. Whereas the already existing atoms for this new molecule to form still have their own "soul"-individuality, there is now a new level of higher intensity energy, which one could call "emergence" and which corresponds to the new "molecular soul entity". The different molecules from which these atoms originated also could have had different molecular entity souls, which were different from the atomic souls of which these were built.

In this way my panpsychic hypothesis diverges from the classical micro(pan)psychism theories, which assume that our consciousness is just the sum of the consciousnesses of the constituents. There is no chicken and egg problem in my hypothesis of what comes first, the meta-system soul or the sub-system souls. The first time a new aggregate is formed it does not have the meta-system soul yet. Such a first-time aggregation is a mutual exploration, not yet a willed morphogenesis. After the constituents "die" and return to their photon state and merge or concatenate because of their convergent development, the new level soul is formed, which can now inhabit newly formed similar aggregates.

The merged meta-system souls now have acquired an energetic morphogenetic blueprint (an energetic interference pattern, which might correspond to what is called an "Astral body" in esoteric circles), which operates as a force field and makes it easier for new aggregates of this level to form. Perhaps the newly formed aggregates do not have the meta-system soul yet, but as they morphologically and energy-wise match the needs of the new higher intensity soul-vector, this new soul-vector can "reincarnate" in this level. Alternatively, the new energy field actively guides the morphogenesis of the aggregate for its incarnation. So in fact you are an aggregate of many levels of souls under the guidance of a highest meta-level soul or consciousness energy string, the "Anima" with which you identify. The subsouls are like the smaller "Animai" in Lucretius' philosophy[91] or like the midichlorians in the film series "Star Wars" –giving you access to Chi or "the Force".

I am aware that this is a highly speculative scheme and I don't sell it as a fact, but it nicely fits a fractal pattern and Tantric philosophy[89]. The Buddha also told that he had taken many different animal life forms in his past lives[92].

From the human form onwards, we may evolve further to become demigods as in the Vedic tradition, a planet or a Sun. To fall back to a human form is also possible. It seems however unlikely to me that one may incarnate in an alien life form, because one's energetic blueprint has no information to match these forms.

It is however not until "Kaivalya" is reached that independence from the "Akashic substrate" is obtained and the "soul" is finally allowed to dissolve in the higher level consciousness it came from. The higher consciousness may integrate the experience of this part of itself, which was nothing more than a game-avatar or a dream from its perspective.

Once many informational scenarios have been lived, embodied and "realised" and the apparent individual soul realises that there is nothing out there but consciousness, it may become quiet and lose the urge for ever ongoing further meta-vari(eg)ation. The Atman might actually wish to undergo a real "final death", tired of its cycles of life and death and having exhausted the passion for repetition. Even if this wave of consciousness, the apparent individual soul, Atman, subsides, the underlying hypostatic consciousness the Parabrahman can never disappear and new waves can be born out of it giving rise to new souls and a new universe. Alternatively, it ends up in a bliss, which turns into a state like deep sleep, the dreams in which create the parallel universes as related by Sri Shankaracharya[93]: "*From the experience of bliss for a long time, there arose in the Supreme Self a certain state like deep sleep. From that (state) māyā (or the illusive power of the Supreme Self) was born just as a dream arises in sleep.*"
Aleister Crowley[78] once said that "every man and every woman is a star" and perhaps we'll all end up as the consciousness of a star one day. Buckminster Fuller[2] describes the intersection of three energy lines as a "star", which in his "geometry of thought" gives rise to "considerations" ("sidera" are "stars" in Latin). But we may even rise further: Eventually each one of us may one day be a Brahma and create

or be his own universe, perhaps we have already been in that state, if our Big Bang was preceded by a transfinite amount thereof before.

The only true "infinity" is the primordial Consciousness, from which everything and all arises and to which everything and all return. It is infinite in the sense that it is "not finite", it has no boundaries. Although it is not limited to any form or structure it can divide out of itself any form and any structure. This does not mean that it necessarily will test out all possible configurations of all possible universes (which is a project without end, since there is no need for a limit in size or number of divisions); perhaps as I already suggested earlier it prunes away scenarios that get stuck in a loop.

Whatever may happen at any given instance, the spatial manifestational dimensions are probably limited and finite: It seems that there can be no spatial infinity or an infinity of dimensions at a given moment, as space and dimensions merely originate from consciousness, but are not identical to it. Originating implies a temporal aspect. Thus, there can be a "temporal transfinity" however, if the HT does not screen and prune and wastes its creative resources.

If the HT encompasses all aspects of existence and allows all these aspects to occur, this means that then any manifestational aspect is also an intrinsic essential characteristic of consciousness in a certain way: It could not have been what it is, if it would have left out one of these aspects. This leads to a "transcending of morality" of the HT which is not within the grasp of many humans: Even if you may not like it, murder, rape, violence, theft, lies etc. are thought forms that have been allowed to arise in the Mind of HT albeit perhaps not in an active form at the highest level, but as an aberration at a lower level. You can blame God (if you want to call the HT that way) for being immoral and cruel, but in fact as every manifestation is ultimately "virtual" and has no real eternal character, it should not harm or offend you. When you have an uncanny unpleasant dream, you'd better get over it, because your dream doesn't affect the outside world you perceive as your "reality". Your dream is even less real than this VR.

Therefore in the HT "morality" is transcended. The HT is unlikely to be a moral bigot who prescribes you rules you are not comfortable with. The HT probably rather gives you the fullest freedom within the chreodes of physical rules to explore every possible way you think might make you happy.

If you pay attention, you will recognise patterns and realise that certain actions you perform, ultimately get you in trouble and make you less happy. Those you can label as unproductive and decide not to repeat them anymore. As everything you object to does happen in HT's simulation there is no point in calling events, things or concepts ultimately "good" or "bad".

It is here that the Indian philosophy of the "kleshas"[23] kicks in. According to Wikipedia the "kleshas" are "mental states that cloud the mind and manifest in unwholesome actions such as fear, anger, jealousy, desire, depression etc." In fact any action carried out in pursuit of a materialistic or egoistic result is said to result in kleshas. Kleshas are not only called "afflictions" but also "visitations", as if you are possessed and haunted, when something unpleasant happens to you. Any action that harms another is believed to return to you via "Karma". The "harmed entity" will take revenge in some form – even after many lives. Perhaps that is why many Indian people are so extreme in practising vegetarianism because they believe that for any entity you kill and eat, you will undergo the consequences of that.

It is probably true that there is some karmic effect in the form of digestive problems, but if you reason this way you make it very difficult for yourself to exist at all. You have the right to exist to and if you must go through the human form, you must eat.

Perhaps by eating a shrimp you will be stung by its next incarnation as a mosquito. I hope the karmic effects won't go much further than that when you eat simple creatures. Eating big mammals like cows, sheep and pigs might not be so harmless. As life forms close to ours, perhaps in a future incarnation, they might come back as human beings and actually try to kill you, or you might come back as a cow and be eaten by them in their future human incarnation. You never know. I don't have a clue how these things work, but it's worthwhile prospecting the notions of Kleshas and Karma in conjunction with reincarnation.

If the HT meant you to exist to reach Kaivalya or Mukti, it also envisaged you being able to eat.

The "visitations" could also be in the form of visitations by ghosts if such entities are possible at all. Perhaps they influence your mental content to go in unpleasant directions. As Terrence McKenna[69] once said: "half of the time you think that you are thinking, you are actually listening".

The philosophy of the "Kleshas" considers that this phase of material existence as in this VR is essentially an unpleasant phase you can better get over with as soon as possible.

It is my personal opinion, that every unpleasant experience holds a great potential for emotional growth. If you consider an experience as unpleasant, this probably indicates that you haven't accepted this part of existence, this part of reality. You try to exclude it. Consciousness as primordial nature from which everything is born however, I suppose to be all-inclusive.
The attitude of labelling an experience as unpleasant is likely largely a mental bias. If you can overcome this bias and surrender your opposition to certain experiences, you might actually learn to accept it, so that it does not cause any emotional distress. You might even learn to enjoy it.

All the so-called "bad experiences" may well be a means to bootstrap your intellect to a higher more encompassing and more compassionate level. Even physical pain need not be "bad". Rather it is likely a warning signal from your body that you must undertake action to resolve a problem. A problem is like a tangled-up knot or a clogged tube. Once you untie the knot or unclog the tube the energy can flow freely again and feel liberated and unbounded.

As Sadhguru Jaggi Vasudev[94] says, any desire is the desire to be limitless. There is nothing bad about desires as long as they don't turn into compulsions, which cause you mental or emotional distress.

Since any compound process in Kaufman's[4] Unified Reality Theory (URT) is also a kind of knot, once could say that the whole informational process in the Akashic matrix can be said to consist of knots (i.e. material objects such as stars, planets, creatures, molecules, atoms) on the one hand and a free information flow in the form of light and other electromagnetic radiation on the other hand. Light that gets trapped in a material knot may need to go through the cycles of life, birth and evolution to escape again.

In the HT all opposites thus find a coexistence accommodation without contradiction. In order to transcend a phenomenon or experience, you cannot exclude it, because than you are stuck with the complementary or opposite phenomenon. The phenomenon or experience needs to be embedded as a part of a higher dimensional entity or process, like the parts of the elephant in the parable of the blind men.
Transcendence in my philosophy involves the generation of ever higher meta-level dimensions to reunite seemingly opposed mutually exclusive phenomena. In analogy hypercomplex numbers become more and more commutative as regards the possible operations going from complex numbers to quaternions,octonions, sedenions etc.

Any situation of apparent "loss" in the end will make you stronger and result in a win-win situation if you can transcend it.

I would like to add a non-exhaustive list of certain apparent contradictions or opposites, which are transcended in the HT:

Aristotelian essences (Aristotle believed every single thing and every phenomenon had its own idiosyncratic essence) vs. the contingency theory of "dependent arising" in Buddhism: In the HT there is an essence to every manifestation, namely consciousness, but consciousness is not an idiosyncratic essence for each phenomenon: It is reductively the same process for every manifestation. As all manifestations arise as dualities, opposites from the same primordial consciousness, they are bound to be linked and contingent on each other, the link being primordial consciousness. We could speak of the contingent essence and the essential contingency.

Creation and evolution do not exclude each other: Within the chreodes of creational patternism, microsingularities are allowed to evolve by aggregation with a certain degree of self-determinacy. The self-creative evolution of the evolutionary creation.

Logic has absurd self-reflexive contradictions, whereas the absurd can be used in logic to reason via the technique of "reductio ad absurdum". Whenever logic starts to inquire about the whole, about the self-reflexive process called consciousness, it fails to operate correctly, as it can only reason with parts. Yet from the absurdities arising we do get insight in the process of consciousness. The self-absurdifying logic and the self-logifying absurdity.

We could establish a rule for naming transcendent concepts: A self-antithetifying thesis.

Opposites are reductively the same, made of the same underlying hypostatic consciousness. Opposites are only apparently different in a relation, but they are identical in intrinsic sense. That's perhaps why Spinbitz[18] speaks of the identity of opposites".

The local is embedded in the global, but what is global from one perspective (like the Earth) can be local from another perspective (e.g. the Galaxy).

Nature accommodates both asexual and sexual reproduction. Hate is a form of love, for if you were truly indifferent to the person in question there could be no hate.

Religion versus Atheism is resolved by the notion of Pancomputational Panpsychism. Panpsychism and Pancomputationalism do not exclude each other but are different sides of the same coin.

As everything is reductively the same, differences are just configurations of sameness.

Determination versus chance or random versus caused are different degrees of different levels as explained before.

Even manifest versus unmanifest is just a matter of perspective depending on the "considerable set" of data that can be maximally apprehended : What is unmanifest to us, can be manifest to other entities (like animals), such as ultrasound or ultraviolet or infrared light.

We have seen that "real" and "virtual" can be a matter of perspective and temporal stability as a set of nested ancestor simulations.

Whatever is expressed in parallel can be expressed in serial and vice versa.

Freedom can also depend on a perspective: Even if an energy is free from being tangled up in a material knot it can still be bound in the Akashic matrix.

Form exists by the grace of the formless void around it. The material is embedded in the spiritual. Order requires a surrounding complementary chaos and vice versa. The ocean requires the drop, the drop the ocean. The complete is a subset of the incomplete. Direct or indirect representations find their absolution in universal representations. The finite is embedded in the infinite. Semantic truth and falsehood are perspectives depending on the amount of information available.

If needed the HT can accommodate all physical and conceptual opposites by nesting the thesis in its antithetic opposite or vice versa. It is a game of levels or layers. A metaphorical kaleidoscopic onion in which what you see depends on the level you are looking at and your considerable set. The numerical ratio between the levels provides for the experience of the Logos: A quality tasted by underlying levels of quantities, which themselves derive from primordial sense. Ultimately "Sense" is primary and structure is deconstructed sense of threefold aggregates that have become definite, condensed and finalised in the form of microsingularities.

As already argued before, because of the great number of numerical and pattern coincidences in the solar system, it is improbable that we do not live in a simulation. If we accept that we live in a simulation, then there must be a highest level, a most advanced simulator level still alive and hence then there must be a HT.

As the HT can make a multiplicity of small islanded consciousness experiences, in a certain sense the HT is relatively omnipotent from within all its simulated VR scenarios.
The very way by which the consciousness of the HT comes to know itself is by its virtualisation, just like your brain serves you a virtualised image and experience of what actually may be going on. In order to predict the outcome of applying a pattern or a substitution thereof, intelligence needs to be able to virtualise scenarios.

It might even be possible that a soul vector can go through exactly the same live as another one, by travelling through its Akashic track. Everything you experienced might be re-experienced by another soul vector, since perhaps all free will is just an illusion and "living" an experience is more like looking to a film.
Or perhaps both are true: There is a first genuine experience by a soul vector who has a degree of free will, but all souls that follow that track after this one do not have a free will and merely watch/live a precooked full immersion experience. Perhaps all experience tracks are already laid down and is life just a repetition of going through tracks, in which case there are just parallel scenarios and cosmic time does not exist.

This corresponds to the Samkhya[95] view of Purusha (the Cosmic Man as metaphor for Consciousness) who merely observes Prakrti (Nature). Perhaps time does exist at the cosmic level and scenarios keep on being developed –which is my preferred interpretation- whilst not excluding the possibility to relive someone else's experience by going through his/her Akashic track.

The HT is most likely not a "moral dude", but probably intelligently screens and prunes away scenarios that have no chance of escaping from a loop.

There are no opposites, which the HT cannot resolve by making additional levels or parallelities. The only thing it may not be able to generate, is its own hypostasis or its own frame of reference, as it is and remains embedded in the impersonal all-encompassing primordial consciousness.

All religions may find their apotheosis in one of the parallel universes, but not necessarily this one. They are hereby incorporated by reference, with the proviso that if any conflict arises between their teachings and Technovedanta, Technovedanta prevails. This way it can be said that all ways lead to "God" without contradiction. I hope your Soul information vector may migrate to your universe of choice upon your death, so that you may find the apotheosis of your own perspective to be "true". All is true to some extent, false to another extent, true and false simultaneously to some extent or neither true nor false to some extent. And if the HT is not true yet, we can make it become a reality.

Thus as Haldane[96] and Terrence McKenna said: "The universe is not only weirder than we imagine, but even weirder than we possibly can imagine". Chaire HT!

## Chapter 10 Transcending Transcendence

There are numerous accounts of people who have had mystical or psychedelic experiences, which went far beyond our normal perception. The ecstatic states described may cause a longing in the interested reader. The question however remains, whether this brings more fulfilment to your life, whether it makes you really happier.
Even if you were to reach the ultimate state of what Buddha calls Nirvana, would life still be worthwhile and if not, is that state of Nirvana really worthwhile?

If we may believe the accounts of the inventor Ithzak Bentov[5], the answer is, yes life is still worthwhile and no, getting stuck in Nirvana deprives you from possibilities of further evolving.

In his book "A Brief Tour Of Higher Consciousness" Bentov[5] described his mystical experiences. He explains how your universe in a state of deep meditation can expand to encompass the whole universe. But he does even go further: Multiple universes appear strung together as a chain of pearls, which form a double helix, like a DNA molecule, which is presided by a radiating Aleph. Did he reach the ultimate Godhead there? No! This Aleph was part of a much larger meta-universal aggregate, which was like a cell. Bentov kept expanding his consciousness and went through various meta-meta-forms of universal aggregates until the ultimate form was a flower and then on through various evolutionary stages until the multi-super-meta aggregate was again the form of a human! In other words Bentov experienced in his journey through meta-levels of manifestations, that existence is a kind of fractal, in which every form gets repeated after a number of levels. Remember that I jokingly said in Technovedanta 1.0[1] that we are perhaps living in the toe of a giant. So even if you evolve to become some kind of cosmic consciousness, if you keep evolving you end up right where you started.

This reminded me of the tale of Chinese stonecutter: A Chinese stonecutter is dissatisfied with his life and is envious of a wealthy merchant. He wishes that he could be like him. To his surprise he changes into the merchant and lives a luxurious life. However, when he sees the king passing by with his entourage, he grows envious of the greater wealth and power of the king and wishes to be like him. Again he is transformed into a new form and now lives as the king. But as a king he grows envious of the power and magnificent radiance and splendour of the Sun and wishes to be the Sun. Again his wish is fulfilled and he becomes the Sun. But as some clouds drift by his rays can no longer touch the Earth and he realises that the power of the clouds exceeds that of the Sun. Therefore he wishes to become a cloud. As a powerful cloud he rains on the Earth causing great mud streams and devastation. But a rock withstands him. He grows envious of the powerful rock and wishes to be a rock. Now as a rock he thinks no one is more powerful than he. Until one day a stonecutter passes by...

What I realised is that in whatever position you are, there is always a position that seems more rewarding or evolutionary more advanced. However, ultimately in those positions one is not better off than you. There is always someone to be envious of. The Siva Samhita[97] mentions that in all levels of existence, even in the heavens, there is suffering due to this type of jealousy. If you continue through the cycles, you may actually end up from where you started. Consciousness in its further expanded forms seems like an unending tower of turtles.

Is there no way out? There is the Nirvana, which Bentov[5] also encountered on his mystical journeys, but this seems to be a place where Souls go to, when they are tired of evolution and do not wish to evolve further. There is bliss, but it seems monotonous and boring in the end.

The whole business of the singularity and transhumanism these days seems to revolve around dissatisfaction and jealousy. If you read articles by Peter Diamandis[98], one of the prominent figures in the Singularity movement, it is all about how to make even more profit for your business. Transhumanists pathetically pursue immortality in the form of "Eugenics", but to me it seems like a hopeless path[99]. Not that we may not be able to prolong our lives substantially via technology, but true immortality in this carbon based substrate seems like an idle wish to me. For that we'll probably need to enter an inorganic silicon matrix of e.g. the World Wide Web.

Whether that is truly a blissful state I dare not say. What I observe among children who grow up in a heavily computerised society and who are hooked to their computer games 24/7, preferably online, is that they have difficulty in performing in the "real" world, the tangible world of everyday life. Their attention span and concentration on non-computerised activities is rather limited. This has been documented in various scientific studies and has been linked to the so-called attention deficit disorder[100]. As more attention and concentration –as I know from my meditational experiences- bring me more blissful feelings, I regret that future generations may not be able to get that experience.

Perhaps I am too pessimistic and perhaps the meditational wheel will be reinvented within the new technological framework, but at present the effects seem to be negative as regards enhancing fulfilment in life.
Another caveat comes from the fact that wherever there is electromagnetic radiation from our electronic devices, this also influences the ability to meditate. As electromagnetic information streams can influence our mind strongly, it is important to seek an equilibrium and have at least part of the day the devices switched off to relax into an emptier void.

So when I am advocating the miracles of Technology, I also think that we must try to alternate our plugged-in states with unplugged states. To have the best from both worlds. The purpose of technology must be to enhance the quality of our lives. Not to create additional problems or psychological states of mental unrest or a numbed down attitude as I see so often with younger people.

I do not wish to write a book about the dangers of technology, but a few encouraging words directed to the fact that in my humble opinion we should strive to develop technology in a responsible and all-inclusive way are in place.

I do this because you may have gotten the impression that I am a technology aficionado. As a patent examiner dealing with inventions on a daily basis I am certainly enchanted by the miracles of human intellect. But I also realise the risks of blind technology development-not for the sake of improving the quality of our lives-but for the sake of making financial profit only. Technology should be a way to unite people of all kinds, not to create islands of exclusivity for the extreme wealthy. Fortunately, history also teaches us that rapid technological development also leads to a more rapid democratisation of technology. The way forward towards a Technological Singularity must be accompanied by an attitude of all-inclusiveness.

One of these risks I do want to mention, is not the dangers of AI or genetics (many books have been written about that), but about the N of nanotechnology in Kurzweil's GNR.

Nanotoxicology is a rapidly developing field these days, but unfortunately too many scientists working with nanoparticles on a daily basis are too unfamiliar with the carcinogenic potential and environmental dangers of these substances[101].

If we truly wish to achieve a technological singularity, we must proceed responsibly and seek for sustainable solutions, for otherwise we may not be there to live it. If we wish to transcend our present condition, we must lay the proper foundation, not only physically but also mentally by transcending our transcendence and appreciate fully the beauty of our current state.

In this way I differ to a great extent from the average Singularitarian, who condemns everything short of that "Singularity" and seems in a constant strive for more and better. The Vedic philosophy of Holism, which states that the drop is not only in the ocean but also the ocean in the drop, makes clear that each one of us has implicitly all the information to unfold every aspect of existence. We are holons in a fractal, and if you go on the inside journey you will encounter the whole universe. This is a very encouraging perspective to me, leaving me fully satisfied with my present condition: I am as complete as I can be and striving for more completeness or striving to transcend my condition is likely a kind of self-delusion, if this becomes a compulsive obsession. Not that I don't like developing myself. I experience a great joy whenever I encounter new novelties in life, I like experiencing new avenues of insights and discoveries. But I can have the tranquillity of mind that there is no necessity to go there. It is my just playfulness, not a compulsion to evolve and transcend beyond transcendence – which is a booby trap.

As regards each and every assertion I made in this book, rest aware and assured that they all remain speculative, but that they might be a useful guideline in our considerable set.

And remember "Everything is hereby incorporated by reference".

## Appendix 1: Technovedanta's answers to the traditional questions of Metaphysics

In chapter 1 of part 1 I promised to answer the traditional questions of Metaphysics from the perspective of the Technovedantic meta-philosophy. Now that you have gained a solid understanding of the most basic concepts of Technovedanta, this chapter is more like a kind of summary, to give you a clear overview. Note that this is a Theory, which I consider as the most likely springboard, but not a conviction or a belief.

### Do I exist?

In order to be able to answer this question, I first have to repeat what is understood by "existence" and "I" in Technovedanta.
In Technovedanta everything that stands out (ex-sists) from the otherwise homogeneous background of primordial consciousness is considered to exist. Existence occurs when energy becomes self-involved and forms a resonant string-loop or in 3D terms a toroid. Energy has thus become been individualised and embedded as wave-particle in the greater metaphorical energy ocean of primordial consciousness. This individualisation accompanied by a feedback which results in a minute form of self-awareness, makes that every sustainable energy quantum thus formed has a sense of subjectivity. Every quantum is a kind of "I" or Purusha. After several stages of evolution such an energy quantum has grown in energy content and has possibly merged with other energy quanta and has become a kind of substrate independent energy, an individual soul or an Atman. The very essence of your experience of subjectivity is this Atman, this "I" or so-called Ego. Each "I" is considered to exist, as it stands out from the homogeneous background of primordial consciousness.

In other words, according to the Technovedantic philosophy "existence" only exists by the very virtue of individualisation of "I"-entities.

However, ultimately in the Technovedantic lore existence is a transient process. Once all "I"-s have attained self-realisation and have merged to become one single entity again, the subsistent primordial consciousness is all that is left and as nothing stands out of it anymore, existence is ended. The subsistent primordial consciousness however remains and can give birth to new cycles of existence if desired.

According to Douglas Adams' *The Hitchhiker's Guide to the Galaxy:* "There is a theory which states that if ever anybody discovers exactly what the universe is for and why it is here, it will instantly disappear and be replaced by something even more bizarre and inexplicable. There is another theory which states that this has already happened"

One could call "existence" illusory in the sense that it is ultimately not perpetual, but from our perspective that would not be fair as we experience existence as rather persistent. Einstein once said: "Reality is an illusion, albeit a persistent one".

Existence is a kind of "consensus-reality" as Kastrup[102] would call it. It is the interference pattern of interactions of all panpsychic entities, which build the material world we see around us. Since material manifestations are the result of "compound-processes" (see chapter 1 of part 2), manifestations can also be considered to be "co-dependently arising" in line with the teachings from Buddhism. Because of this co-dependency and the fact that each Purusha or soul has a degree of free will, the system is not pre-determined.

The materialist stance is that we don't really know, whether we exist since even Rene Descartes' "cogito ergo sum" (I think therefore I am) can be based on an illusion generated by the material interactions of the brain. But that would still acknowledge that brains exist and that matter exists –even as transient quantum fluctuation.

René Descartes walks into a bar. The bartender asks if he wants anything. … René says, "I think not," and then "poof" magically disappears into nothing.

Even if our experience of subjectivity would be engendered by the brain or by a virtual reality simulation in which we live, for us it does not matter. As long as we experience anything as not being ourselves, as long as we exclude any perceptory content from our I-ness, we will experience subjectivity. Even if this sense of subjectivity is all there is to I-ness, it is enough to demarcate it as an "I". This I-ness is an experiential reality that cannot be denied in any way. Even if the world around us would be an imagination of this "I"-ness, this "I" would experientially sufficiently be distinguished from its other experiential content, to conclude that at least "I" exist and by that virtue establish an existence as such. By definition I can then state that according to Technovedanta I exist.

To conclude that you are a philosophical zombie or a Ghost would in no way correspond to your "experiential reality", although it is true that there are disorders that lead to a certain degree of self-alienation accompanied with a sense of not belonging to your body. In Cotard's syndrome a person holds the belief that he or she does not exist or is already dead. As long as the apparent outside world still can influence you and vice versa, you are still part of that reality, that existence and you cannot conclude that you do not exist. The mere fact that these patients can relate what they experience proves (and should prove to them) that they still belong to this consensus-reality (even if it is a virtual reality). However, they may think that everything they experience is made up by their minds. Funny or sad enough, mentally ill people often laugh when at they are at a point where they cannot distinguish reality from their delusions.

The body cannot be "You", as its material constituents are continuously exchanged with the environment. The brain is not "You" either, since it has turned out to be much more plastic than previously believed and mental disorders can sometimes be completely cured, resulting in extensive rewiring of the neurons.

I am not my body I am not my thoughts, still I am this sense of subjectivity, which is unchanged as long as I know myself.

**How do I know I exist?**

To be able to answer this question, we first have to establish what is understood by "knowing". Knowledge consists of pieces of information, that have been understood, grounded and abstracted from a vast amount of data gathered by experience, including education, and have been stored in our Mind as a potentially useful memory that can be called upon, if a similar situation arises.

The repeated experience of subjectivity leads to the experiential knowledge that there is a part of our experience which we consider to be sufficiently distinguished from its other experiential content. What we can feel from the inside, we consider as part of our "I". What we can only experience with our outward oriented senses, we consider as belonging to the rest of existence, the "outside world".

We know our existence thus experientially. But the borders of this I-ness need not remain the same. People with prostheses may develop a sense of subjectivity even in their prosthetic parts. Meditation can result in experiences of becoming one with the object of meditation and experiencing that object as if one is feeling it from the inside. One embodies an object previously deemed to be outside one self. By doing so, one starts to merge with reality, one starts to include more of reality and to expand the sense of I-ness. Conversely a sense of disembodiment of one's own body as in Cotard's syndrome can be disconcerting. Only a scientific attitude can then be of help (even if one considers that science has its limits). Experiments for such a person can be to test whether repeated actions X result in the same effect Y in what is considered the outside world. The stimulation of areas of the brain network which are normally associated with internal awareness can be the physical treatment for such a disorder.

**What is consciousness?**

This book has described primordial consciousness as the foundation of both existence and experience.

In this pancomputational panpsychic T.O.E. only primordial consciousness itself is considered as irreducible: It cannot be described in terms of constituents as it is itself the most fundamental dimension of being. Therefore I also sometimes call it "the Absolute", because it is not relative to any concept or object. Rather, all objects and concepts are made from primordial consciousness. I do not wish to call it the "*prima materia*", because it is not material. Rather materiality emerges from smaller self-reflective loops, vortexes in the greater primordial consciousness.

Primordial consciousness is however not to be confused with animal or human consciousness, the awareness of an object in the mind or another definition in terms of something else. Whereas these forms of consciousness are very direct derivatives of consciousness and reflect the way primordial consciousness expresses itself via us, they are not directly the primordial consciousness I am discussing.

Whereas primordial consciousness cannot be described or defined in terms of constituents, it can however be illustrated in terms of its inextricable intrinsic aspects: Its behaviour and way of manifesting itself.

Primordial consciousness is not a thing or phenomenon; it is rather a process: It is essentially a self-reflective feedback loop, which has the qualities of experiencing subjective being. In material terms one would consider this primordial consciousness a kind of "nothingness", a void. This is also how the Buddhists call it, Shunya. But it is not nothing, even if it is not a thing. Rather it is neither nothing nor a thing. It is that what senses, what experiences in every being.

In the Indian philosophy of Samkhya it is called "Purusha" (the cosmic man); in the philosophy of Vedanta it is known as the Parabrahman, the highest Brahman, which is beyond all descriptions and conceptualisations. In Western terminologies, the terms which would come closest are perhaps "Being", "God" or the "Soul", yet these terminologies do not fully capture the all-encompassing nature of primordial consciousness, nor am I comfortable with these terminologies. As you know, I am not particularly fond of religions at

all, as they make our understanding of our existence often more cumbersome.
The process how primordial consciousness functions can perhaps best be illustrated by the metaphor of a "Torus" (the form of an apple in abstracted sense). From the central point in the Torus primordial consciousness expands its energies in order to get to know itself, to become self-aware, to sense itself. As the energies rise, they fold back on themselves to return to the central point via the other side; the proverbial arse of the Torus. Once the energies return to the central point, the feedback is complete: the primordial consciousness senses the return of its energies and is updated with the information of the journeys of its energies. It senses itself, it has become self-aware.

Every energy that expands from primordial consciousness, can in a fractal way undergo a process of sensing itself, form a miniature Torus type vortex superimposed on the greater Torus, thereby forming a fractal of consciousness. Additionally, all vortexes or mini-Toruses (subTelors, i.e. souls) can generate energies themselves and sense each other. I do not say however that this "is" so; I merely wish to use a metaphor, which is helpful in order to understand the basics of my T.O.E. So we are like fractalised torus tentacles of the primordial torus.

This is a very different view than the reigning materialistic paradigm which considers consciousness as a trick of the mind, trying to prove to itself that it exists. Decisions are taken by unconscious processes and when they bubble to the surface of the conscious perception, this imaginary sense of consciousness appropriates these findings from the subconscious.

Whereas I do agree with the notion, that a great deal of decisive processing is made subconsciously by autonomous circuits in our minds, I am not willing to accept that all decisions are made by this process. When I am concentrated, I can will myself to ignore the course of action that my subconscious proposes. I can take a sudden adventurous leap in the unknown which is completely contrary to any routine that my basal ganglia try to trigger. This ability to deviate from predetermined pathways is in fact already present at the lowest level of existence we know of, resulting in the Heisenberg uncertainty of the quantum dimension. From this panpsychic stance, free will is an inherent aspect of any conscious entity, including subatomic particles. (This also answers the question do we have free will? I wanted to write something about free will, but decided not to).

Whereas there is no need to suppose that there are countless parallel universes in which each and every possible decision of each and every possible entity is made, there is also no reason to exclude such a scenario a priori, although as explained before it would be a waste of resources.

In imitation of the primordial consciousness, at the smallest level a quantum fluctuating energy can loop into its own pathway and can establish a standing wave, if an integral number of half wavelengths fit in a closed string. It thereby comes to know itself and proto-sense becomes proto-consciousness. Thus a "particle" e.g. a boson or a lepton can be formed, which has a minute form of consciousness. A miniature being or soul has been formed, which at the same time forms a material or energetic particle. This then also answers the question:

**What is reality made of?**

Reality is *inter alia* made of self-revolving individualised vortexes that from our perspective are observed as subatomic and energetic particles. Matter is a compound process, photonic energy and other electromagnetic radiation an uncompounded process of such entities. But as explained before in chapter 1 of part 2, these entities do not "exists" in the absence of a "substrate". Certain self-revolving individualised entities constitute the so called matrix of reality cells, which are nothing more than the simplest relations of primordial consciousness with itself, in which the above mentioned particle/wave forming vortexes can express themselves. If a reality cell is empty its value can be considered as "zero", if it is occupied by an energetic entity its content is "one". Thus the reality cell matrix and its occupants together form a kind of digital checkerboard game, resulting in an inhabited substrate which is both pancomputational and panpsychic.

Don't trust atoms, they make up everything.

**Why is there something rather than nothing?**

This question remains unsolved in physics. If reality were to arise from a true nothingness, there should be as much anti-matter as there is matter, or there should be a counter anti-matter universe which we haven't observed thus far. The quantum vacuum however has not turned out to be a nothing, but rather to be boiling with activity. Still this does not explain yet why this should give rise to a Big Bang.

From the Technovedantic perspective the answer is more likely to be found in the desire to express and know itself of the primordial consciousness. Consciousness can only become aware of itself if it sends out energy to probe itself. Existence is then the process via which the primordial consciousness probes itself in all its manifestations and is the only way via which it can perpetuate itself.

**What is the meaning of life?**

"Meaning" as in "the meaning of life" is often understood as its purpose. In imitation of Langan[3], the purpose of life should be Telesis, the maximisation of utility for all its participants. But the very will to maximise utility for all its participants can be considered as an act of love of the protean Telor, the Aion Teleos or Highest Transcendence for the participants in its game "Leela", to have the participant subTelors or souls experience the joy of getting to know themselves and their ultimate realisation that they are one with the primordial consciousness. Therefore "love" and "enjoyment" are equally valid answers all fitting in the notion of "maximisation of utility for all its participants".

The meaning of "Meaning" in this book was however often explained as "proximity co-occurrence". Information contextually acquires meaning if there is a grounded pattern of proximity co-occurrence of at least two terms, a didensity. Similarly physically particularisation in the form of matter occurs upon proximity co-occurrence, giving rise to Kaufman's "compound processes". This communion of entities in the form of a proximity co-occurrence can also be considered as a form of "love", which occurs as long as the participants can enjoy each other's presence and learn from each other. Thus, also via the meaning of meaning we arrive at the conclusion that mutuality resulting in loving enjoyment is the meaning of life.

**Where do good and evil come from?**

Good and evil are relative terms. Morality from a human perspective depends on your cultural and religious background. However, at the higher level of simulation we live in, we could consider the principle of the "maximisation of utility for all its participants" as the system's moral guideline. From that perspective every act that aligns with and is intended to adhere to this principle could be considered as relatively "good". Deviation from that guideline could be considered as relatively "bad".

Still there can be "good" in the purported evil: The very essence of existence being the illusion of individualisation, it is logic that the subTelors must go through a process of getting to know themselves and realising what the meaning of life is. If this process was too easy, there would be no point in existing: In fact it is only by making mistakes that the subTelors can discover which type of actions does not align with the system's morality. Therefore, it is likely that from the onset of the self-creation of existence / the simulation it was in-built / programmed that deviation from the guideline could occur, yes, it was probably even deemed necessary. There can be no good, if it cannot be contrasted with something which is considered "bad". There can be no evolution, if everything is perfect from the onset.

In fact as "evil" is a necessary requirement of existence, it is ultimately something good. This does not mean that I incite you to kill your neighbour, but I also do not recommend you to live like a saint.
So from a transcendent perspective there is no dichotomy of good and bad. All non-aligning actions of a subTelor are necessary hurdles in its evolution, to bootstrap it to the level where it realises that communion is more rewarding than individualisation. Once the subTelor has completely exhausted its egoistic urges giving rise to its existence, it is ready to return to the source of primordial consciousness.

**What is time?**

The issue of time has been comprehensively explained in chapter 5 of part 2. On the one hand, from inside the simulation matrix time is an emergent property resulting from the compounded dynamics of multiple energy content entities propagating through the energy matrix in compound processes called matter. Without matter there is no time in Kaufman's model. It is the measure of the periodicity of a compound process.
Then there can be a higher level of time, a kind of cosmic time of the entities that simulate our existence. From this level, time quanta in which things happen in our simulated reality can be alternated with time-gaps in which the simulators evaluate and tweak the system. The highest or most fundamental level of time is possibly a measure for the inherent periodic changes in the dynamics of primordial consciousness. Thus time can be expressed as a fractal effect of the AION Teleos with different self-similar periodicities of the subTelors.

**Does God exist?**

What is God? What are Gods? As explained in this book, it is likely that we live in a simulation and our simulators could be considered as a kind of Gods from our perspective. From the numerous numerical coincidences in the solar system, their (former?) presence appears undeniable.

Perhaps they have created a hypercomputer in which organic and artificial intelligence have merged, forming the so-called Eschaton at the end of time. We don't know, but they certainly have great computational powers. They are however not omnipotent or omniscient, although endowed with great powers.

I have also suggested that there is one level of simulators (if there is a series of stacked nested simulations) which is the most developed simulating level, which can be considered as the Highest Transcendence, the AION Teleos or protean Telor. Still this entity or these entities may not necessarily have merged with the Primordial Consciousness.

Is the Primordial Consciousness (PC) then God? The PC is not necessarily simultaneously aware of everything which is happening in its embedded dimensions, but may be a single experience of only bliss. I don't know, since it does not fall within my experience yet. The PC can be considered as being endowed with all qualities since everything is embedded in it, but this does not necessarily mean that it has a simultaneous awareness of all these qualities. Therefore, if you call the PC God, it may not be God as you know it, endowed with the traditional qualities of omniscience, omnipotence and standing outside of time. It rather subsists than exists. It stands outside of our time, but is not necessarily timeless. What it knows and experiences, only the few who have experienced Samadhi or Satori may be able to relate to us, and perhaps not even these.

**References:**

All references are hereby incorporated by reference in their entirety. Wherever their teachings contradict the present teachings, Technovedanta 2.0 prevails.

[1] Tuynman A. "Technovedanta, Internet architecture of a quasiconscious Vedantic Webmind, a panpsychic Theory of Everything", Lulu, 2012.
[2] R. Buckminster Fuller, "Synergetics: Explorations in the Geometry of Thinking", Macmillan, 1982.
[3] C.M. Langan, "The Cognitive-Theoretic Model of the Universe: A New Kind of Reality Theory" Progress in Complexity, Information and Design, 2002.
[4] S.Kaufman "Unified Reality Theory: The Evolution of Existence Into Experience", Destiny Toad Press, 2002.
[5] I.Bentov, "A Brief Tour of Higher Consciousness: A Cosmic Book on the Mechanics of Creation", Destiny Books, 2000.
[6] (http://www.wired.co.uk/news/archive/2015-09/14/darpa-creates-feeling-prosthetic-arm).
[7] Oizumi M, Albantakis L, Tononi G " From the Phenomenology to the Mechanisms of Consciousness: Integrated Information Theory 3.0". PLoS Comput Biol 10(5): e1003588, 2014.
[8] https://terencemckenna.wikispaces.com/Eros+and+the+Eschaton
[9] P.Teilhard de Chardin "The Phenomenon of Man". Harper Collins, 2002.
[10] R.Kurzweil, "The Singularity is Near: When Humans Transcend Biology", Viking Press Inc., 2005.
[11] H.Bloom, "The God Problem", Prometheus Books, 2012.
[12] R. R.Llinás, "I of the Vortex: From Neurons to Self", MIT Press, 2002.
[13] Ben Goertzel, "Creating Internet Intelligence: Wild Computing, Distributed Digital Consciousness, and the Emerging Global Brain" IFSR International Series on Systems Science and Engineering, Vol. 18, Kluwer Academic/Plenum Publishers, 2002.
[14] D.Bohm, "Wholeness and the Implictate order", Routledge, Ed.2002.
[15] R.Abraham, S.Roy, "Demystifying the Akasha: Consciousness and the Quantum Vacuum", Epigraph Publishing, 2010.
[16] T.Campbell, "My Big TOE", Lightning Strike books, 2007.
[17] Verlinde, E.P. arXiv:1001.0785.
[18] J. Morrison, " Volume I - Interface Philosophy, Mathematics, and Nondual-Rational Empiricism", Joel Morrison, 2007
[19] http://cosmometry.net/vector-equilibrium-&-isotropic-vector-matrix
[20] Martineau, J. "A Little book of Coincidence", Wooden Books, 2001.
[21] http://www.ukapologetics.net/09/cumming2.html

[22] A.Wallace, "Hidden Dimensions: The Unification of Physics and Consciousness", Columbia University Press, 2010.

[23] Patanjali, "Patanjali's Yoga Sutra" Penguin Classics, 2009.

[24] M.Heidegger, "Being and Time", Harper Perennial Modern Thought, Ed.2008.

[25] https://en.wikipedia.org/wiki/Russell's_paradox

[26] D.R.Hofstadter "Gödel, Escher,Bach: An eternal golden braid", Penguin Books , 1979.

[27] Rupert Sheldrake, "Morphic Resonance: The Nature of Formative Causation", Park Street Press, 4th Ed. 2009.

[28] Ben Goertzel, "The Hidden Pattern", Brown Walker Press, 2006.

[29] http://www.kurzweilai.net/forums/topic/a-question-for-setai#post-48377

[30] http://arxiv.org/pdf/1401.1219v2.pdf

[31] Bryce Seligman DeWitt, R. Neill Graham, eds, "The Many-Worlds Interpretation of Quantum Mechanics", Princeton Series in Physics, Princeton University Press (1973), Contains Everett's thesis: The Theory of the Universal Wavefunction, pp 3–140.

[32] J.Mitchell, "How the World Is Made: The Story of Creation according to Sacred Geometry", Inner Traditions, 2009.

[33] https://www.discprofile.com/what-is-disc/william-marston/

[34] https://www.linkedin.com/pulse/20130121160547-38251380-ideas-are-more-valuable-than-execution

[35] T.Leary, "Info-Psychology", New Falcon Publications, 1987.

[36] R.A.Wilson, "Prometheus Rising", New Falcon Publications, 1983.

[37] http://www.kurzweilai.net/forums/topic/meta-system-extropy-exploring-a-panpsychic-formulation-of-a-new-measure-of-order

[38] http://arxiv.org/abs/1009.5287

[39] P.K.Dick, "Valis", Mariner Books, 1978.

[40] P.K.Dick, "Exegesis", Houghton Mifflin Harcourt, 2011.

[41] A.Crowley, "The Goetia: The Lesser Key of Solomon the King: Lemegeton - Clavicula Salomonis Regis, Book 1", Red Wheel, 1995.

[42] Hermes Trismegistos, Tabula Smaragdina, available on http://www.sacred-texts.com/alc/emerald.htm

[43] https://twitter.com/setai/status/239397436144095232

[44]S.J.Gates in http://arxiv.org/abs/0806.0051, 2008.

[45] Wheeler, John A. (1990). "Information, physics, quantum: The search for links". In Zurek, Wojciech Hubert. Complexity, Entropy, and the Physics of Information. Redwood City, California: Addison-Wesley.

[46] Bostrom, N., "Are You Living in a Simulation?", Philosophical Quarterly, Vol. 53, No. 211, pp. 243-255, 2003.

[47] Peter Russell, " From Science to God: A Physicist's Journey into the Mystery of Consciousness," New World Library, 2005.
[48] Daniel M. Harris, Julien Moukhtar, Emmanuel Fort, Yves Couder, and John W. M. Bush, "Wavelike statistics from pilot-wave dynamics in a circular corral", Phys. Rev. E 88, 011001(R), 2013.
[49] Kim, Yoon-Ho; R. Yu; S.P. Kulik; Y.H. Shih; Marlan Scully (2000). "A Delayed "Choice" Quantum Eraser". Physical Review Letters. **84**: 1–5.
[50] http://www.azquotes.com/quote/723734
[51] Rupert Sheldrake, "The Science Delusion", Coronet, 2012.
[52] A hypothesis for this phenomenon is the thermal recoil force, see https://en.wikipedia.org/wiki/Pioneer_anomaly
[53] F.J.Tipler, The Physics of Immortality, Anchor Books, 1994.
[54] http://joedubs.com/
[55] http://hubpages.com/education/Number-9
[56] Jan Wicherink, Souls of Distortion Awakening, Self-published, 2006, http://www.soulsofdistortion.nl/SODA_toc.html
[57] Scott Onstott, Tripartite, SIPS Productions Inc, 2015.
[58] Christopher Knight and Alan Butler, "Who built the moon", Watkins Publishing, 2005.
[59] Robin Heath, Sun, Moon and Earth, Wooden Books, 2001.
[60] (http://www.biblewheel.com//GR/GR_37.php).
[61] The Rig Veda: Complete, Forgotten Books, 2008.
[62] A.C. Bhaktivedanta Swami, Srimad Bhagavatam, Bhaktivedanta Book Trust, 2004.
[63] W.J.Moore, "Schrödinger: Life and Thought", Cambridge University Press, 1992.
[64] Sadhguru J.V. "Body - The Greatest Gadget / Mind Is Your Business", Diamond Books, 2013.
[65] John 1:1, The Bible.
[66] Muhammad, The Qur'an.
[67] T.J.Routt, "Quantum Computing, The Vedic fabric of the Digital Universe ", 1st World Library, 2005.
[68] Moses, Torah, Pocket Edition Paperback, Edited by Inc. Jewish Publication Society (Editor), 2000.
[69] http://www.goodreads.com/quotes/518786-half-the-time-you-think-your-thinking-you-re-actually-listening
[70] M. A. S. Abdel Haleem (Translator), "The Qur'an", Oxford University Press, 2008.

[71] Malaclypse The Younger and Omar Khayyam Ravenhurst, "Principia Discordia", Loompanics Unlimited, 1980.

[72] A.Tuynman, "Technovedanta 2.0, Transcendental metaphysics of Pancomputational Panpsychism", Lulu, 2016.

[73] https://herosodysseyphysics.wordpress.com/proofs-of-moving-dimensions-theory-mdt/

[74] http:// beyond-information.blogspot.com

[75] B.Whitworth, https://arxiv.org/ftp/arxiv/papers/1011/1011.3436.pdf

[76] https://cnx.org/contents/rt2I8D90@3/The-Wave-Nature-of-Matter-Caus#import-auto-id2628777

[77] https://www.youtube.com/watch?v=BEColjHXkVQ

[78] Aleister Crowley, "Magick: Liber ABA Bk.4," Red Wheel/Weiser; 2nd Rev. Ed., 1998.

[79] Terrence McKenna, "Food Of The Gods: The Search for the Original Tree of Knowledge", Rider & Co, 1999.

[80] https://quadriformisratio.wordpress.com/2013/07/01/quadriformis-ratio/

[81] L.Wittgenstein, "Tractatus Logico-Philosophicus", Cosimo Classics, Ed. 2007.

[82] https://www.youtube.com/watch?v=RChE0hDRNLs

[83] V.Schauberger, "The Energy Evolution: Harnessing Free Energy From Nature", Gill & MacMillan, 2000.

[84] D.Winter, "Implosion's GRAND ATTRACTOR", 2011.

[85] Tim Gross https://twitter.com/setai/status/405063788488970240

[86] Nagarjuna, "The Fundamental Wisdom of the Middle Way: Nagarjuna's Mulamadhyamakakarika", Oxford University Press, 1995.

[87] http://www.kurzweilai.net/forums/topic/recent-hard-takeoff-rants-9-of-9-the-new-worldview-a-simulist-ouroboros

[88] http://isha.sadhguru.org/blog/yoga-meditation/science-of-yoga/beyond-big-bang-science-spirituality/

[89] Gavin and Yvonne Frost, Tantric Yoga, Motilal Banarsidass Publishers, 1989.

[90] I.K.Taimni, Man, God and the Universe, A Quest Book, 1969.

[91] Lucretius, "De Rerum Natura," Clarendon Press, 2nd Ed. 1963.

[92] Thich Nhat Hanh, Old Path White Clouds, walking in the footsteps of the Buddha, Parallax Press, 1991.

[93] http://veda.wikidot.com/maya

[94] Sadhguru J.V., " Essential Wisdom From A Spiritual Master", Jaico Publishing House, 2008.
[95] Ellwood Austin Welden, "The Samkhya Karikas of Is'vara Krishna with the Commentary of Gaudapada", BiblioBazaar, LLC, 2009.
[96] https://en.wikiquote.org/wiki/J._B._S._Haldane
[97] James Mallinson, "The Shiva Samhita", YogaVidya.com, 2007.
[98] http://singularityhub.com/2015/05/11/the-world-in-2025-8-predictions-for-the-next-10-years/
[99] https://en.wikipedia.org/wiki/Transhumanism
[100] P.A.Chan and T.Rabinowitz, Annals of General Psychiatry 2006, **5**:16.
[101] Tsuda H, et al. Toxicology of engineered nanomaterials - a review of carcinogenic potential, Asian Pac J Cancer Prev. 2009;10(6):975-80.
[102] B.Kastrup, "Why Materialism is Baloney", iff Books, 2013.

www.ingramcontent.com/pod-product-compliance
Lightning Source LLC
Chambersburg PA
CBHW060824170526
45158CB00001B/78
* 9 7 8 1 3 2 6 8 9 3 1 2 5 *